Fundamentals of Computational Methods for Engineers

Edited by

Md. Masud Rana

Department of Electrical & Electronic Engineering
Rajshahi University of Engineering & Technology, Kazla-6204
Rajshahi, Bangladesh

Wei Xu

Professor and Senior Member, IEEE,
School of Electrical and Electronics Engineering
Huazhong University of Science and Technology (HUST)
China

&

Youguang Guo

School of Electrical and Data Engineering
University of Technology Sydney (UTS)
Sydney, Australia

Fundamentals of Computational Methods for Engineers

Editors: Md. Masud Rana, Wei Xu & Youguang Guo

ISBN (Online): 978-981-5039-05-4

ISBN (Print): 978-981-5039-06-1

ISBN (Paperback): 978-981-5039-07-8

©2022, Bentham Books imprint.

Published by Bentham Science Publishers Pte. Ltd. Singapore. All Rights Reserved.

need for a court order if at any point you breach any terms of this License Agreement. In no event will any delay or failure by Bentham Science Publishers in enforcing your compliance with this License Agreement constitute a waiver of any of its rights.

3. You acknowledge that you have read this License Agreement, and agree to be bound by its terms and conditions. To the extent that any other terms and conditions presented on any website of Bentham Science Publishers conflict with, or are inconsistent with, the terms and conditions set out in this License Agreement, you acknowledge that the terms and conditions set out in this License Agreement shall prevail.

Bentham Science Publishers Pte. Ltd.
80 Robinson Road #02-00
Singapore 068898
Singapore
Email: subscriptions@benthamscience.net

BENTHAM SCIENCE

CONTENTS

PREFACE

The proposed textbook bridges the gap between the advanced and the introductory books in numerical methods for engineering, particularly electrical engineering. Although various outstanding texts cover in detail the introductory topics suitable for undergraduates but many topics have been ignored to higher level undergraduate and the graduate. These topics include error analysis, algorithm of the methods, the application of numerical methods for engineering problem, the introduction of higher lever numerical methods such as finite difference time domain method and method of moments. The book exposes these topics in sufficient detail while retaining the usual introductory topics in numerical methods, which makes it useful textbook for both undergraduate and introductory graduate courses in computational methods.

CONSENT FOR PUBLICATION

Not applicable.

CONFLICT OF INTEREST

The author declares no conflict of interest, financial or otherwise.

ACKNOWLEDGEMENTS

It is a pleasure to thanks Ms. Humaira Hashmi, Editorial Manager of Publications of the eBooks Publication Department for her valuable assistance. My love and gratitude to my wife Hafiza and children Mehjabin Maimuna, Sadiidah Munira, and Abdullah Bin Masud for their support, help, and encouragement.

Md. Masud Rana
Department of Electrical & Electronic Engineering
University of Engineering & Technology, Kazla-6204
Rajshahi
Bangladesh

Wei Xu
Professor and Senior Member, IEEE, School of Electrical and Electronics
Engineering
Huazhong University of Science and Technology (HUST)
China

&

Youguang Guo
School of Electrical and Data Engineering
University of Technology Sydney (UTS)
Sydney
Australia

List of Contributors

Mahabubur Rahman   Department of Electrical & Computer Engineering, Rajshahi University of Engineering & Technology, Bangladesh

Masud Rana   Department of Electrical & Electronic Engineering, Rajshahi University of Engineering & Technology, Bangladesh

Rashidul Islam   Department of Electrical & Computer Engineering, Rajshahi University of Engineering & Technology, Bangladesh

Shamim Anower   Department of Electrical & Computer Engineering, Rajshahi University of Engineering & Technology, Bangladesh

Tanvir Ahmed   Department of Electrical & Electronic Engineering, Rajshahi University of Engineering & Technology, Bangladesh

Wei Xu   Department of Electrical & Electronic Engineering, Huazhong University of Science and Technology (HUST), China

Youguang Guo   Department of Electrical & Electronic Engineering, University of Technology (UTS), Sydney, Australia

CHAPTER 1

Introduction of Computational Methods

Md. Masud Rana[1*], Wei Xu[2] and Youguang Guo[3]

[1] *Department of Electrical & Electronic Engineering, Rajshahi University of Engineering & Technology, Bangladesh*

[2] *Department of Electrical & Electronic Engineering, Huazhong University of Science and Technology (HUST), China*

[3] *Department of Electrical & Electronic Engineering, University of Technology (UTS), Sydney, Australia*

Abstract: In this chapter, fundamentals of computational methods are presented. Mathematical modeling of the physical systems is described. Then, various type of methods, error analysis and the algorithm with software packages are also discussed here.

Keywords: Algorithm, Error, Mathematical modeling, Program.

INTRODUCTION

During the pre-computer era, scientists and engineers generally use three different ways for solving mathematical problems: (i) analytical or exact methods, (ii) graphical technique, and (iii) manual calculations [1]. As a result, a significant amount of energy is required on the solution technique rather than on problem definition and interpretation. Today, computational methods and digital computers provide an alternative for such complicated solutions. Computational methods use a number to simulate the mathematical processes, which in turn usually simulate real world situations. There are also several reasons to study the computational methods [2].

(i)   To solve problems that cannot be solved analytically and exactly.
(ii)   To solve problems that are intractable.
(iii)   To reduce the number of prototypes constructed.

---
*Corresponding author Md. Masud Rana:** Department of Electrical & Electronic Engineering, Rajshahi University of Engineering & Technology, Rajshahi, Bangladesh; Tel: +8801724550535; E-mail: md.masud.rana.ruet@gmail.com

(iv)   To use as the forward solver in an optimization routine to gain the best possible design.

Computational methods manipulate ($+/-,\times,\div$, *etc.*) numerical values rather than derive or manipulate analytical mathematic expressions ($\frac{d}{dx},\int dx, e^x, x^b, \ln x$, *etc.*). It always deals with approximate values. However, to obtain efficient results from the computational analysis, one requires realistic mathematical modeling of physical systems. Mathematical modeling is used to numerically study the behavior of complex systems employing a computer simulation.

MATHEMATICAL MODELING OF PHYSICAL SYSTEM

A mathematical model usually describes a system by a set of variables and a set of equations that establish relationships between the variables. It explicates the essential features of the physical system [3]. It also helps to study the effects of different components, and to make predictions about the behavior of the system. The mathematical model can be represented as a functional relationship of the following form:

Dependent variable= function (independent variables, forcing functions,

parameters)   **(1.1)**

where the dependent variable is a characteristic that usually reflects the behavior or state of the system; the independent variables are usually dimensions, such as time and space, along which the system's behavior is being determined; the parameters are reflective of the system's properties or composition; and the forcing functions are external influences acting upon the system [4]. A traditional mathematical model for physical science contains most of the following elements:

1. Governing equations

2. Supplementary sub-models

   i. Defining equations

   ii. Constitutive equations

3. Assumptions and constraints

   i. Initial and boundary conditions

   ii. Classical constraints and kinematic equations

In general, mathematical models may include logical models. In many cases, the quality of a scientific field depends on how well the mathematical models developed on the theoretical side agree with the results of repeatable experiments. Lack of agreement between theoretical mathematical models and experimental measurements often leads to important advances as better theories are developed. Thus, most engineering problem solving employs the two-pronged approach of empiricism and theoretical analysis, as shown in Fig. **(1.1)**.

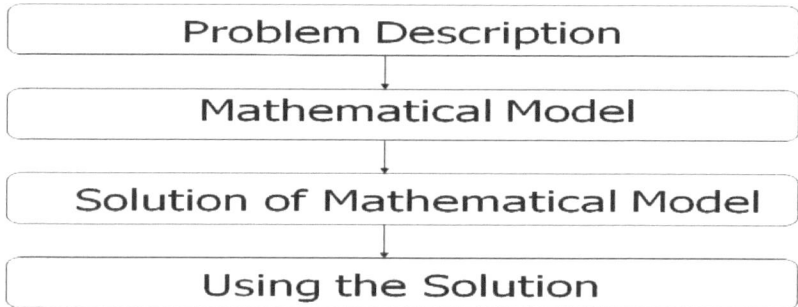

$$\boxed{\text{Problem Description}}$$

$$\boxed{\text{Mathematical Model}}$$

$$\boxed{\text{Solution of Mathematical Model}}$$

$$\boxed{\text{Using the Solution}}$$

Fig. (1.1). Physical problem solving process

The actual mathematical expression of Eq. **(1.1)** can range from a simple algebraic relationship to large, complicated sets of differential equations. For example, Newton formulated his second law of motion, which states that the time rate of change of momentum of a body is equal to the resultant force acting on it. The mathematical expression, or model, of the second law, is the well-known equation.

$$F = ma \, , a = \frac{dv}{dt} = \frac{F}{m} \tag{1.2}$$

where, F net force, m mass of the object, and $\frac{dv}{dt}$ rate of change of velocity. Compared to equ. (1.1), it can be said that $a = \frac{dv}{dt} =$ the dependent variable reflects the system's behavior, $F =$ the forcing function and $m =$ a parameter representing a property of the system. Now, we will explain some mathematical models in the engineering systems. For example, Kirchoff's proposed mathematical model for the electrical network that for each node $\sum \text{current}\,(i) = 0$ and for each loop, $\sum \text{emf's} - \sum \text{voltage drops for elements} = 0$. For civil engineering, force balances are utilized to analyze structures such as the simple truss. At each node of truss, $\sum \text{horizontal forces}\,(F_H) = 0$ and $\sum \text{vertical forces}\,(F_H) = 0$. Similar principles are employed for mechanical engineering applications to obtain the mathematical model to analyze the transient up and down motion or vibration of an automobile. Thus, many engineering applications can be designed to illustrate how numerical methods are employed in the engineering problem-solving process in the rest of the chapter.

COMPUTATIONAL METHODS FOR ENGINEERING SYSTEM

Different type equations that are usually obtained from mathematical modeling of physical systems are nonlinear, linear algebraic, tabulated data, differentiation, integration and differential equations. To solve those equations numerically, various type methods are used, which can be summarized as follows:

i. To find out the roots of nonlinear equations, various methods such as Bisection, Secant, and Newton Raphson, *etc.,* are used. These are described in Chapter 2.

ii. To solve the systems of linear algebraic equations, the direct methods (Gauss elimination, LU decomposition, *etc.*) and indirect numerical methods (Gauss Seidel, Jacobi, *etc.*) are applied. Linear algebraic equations originate from the mathematical modeling of large systems of interconnected elements such as electric circuits, structures, and fluid networks. All the methods are explained in Chapter 3.

iii. To reconstruct a function from the known set of points, tabular points or arguments or to replace a function into tabular data, various interpolation and extrapolation methods such as Lagrange, Newton, Cubic spline methods are used. These methods are discussed in Chapter 4.

iv. To fit curves to data points, various techniques such as regression and interpolations are used, which are described in Chapter 4.

v. To determine the area under a curve, numerical integration techniques such as Trapezoidal, Simpson, Romberg, Weddles *etc.*, are used. On the other hand, Newton's formula is used for numerical differentiation. Chapter 5

vi. To solve the differential equations, various techniques such as Euler, Runge Kutta *etc.* are used. Chapter 6

vii. Advanced numerical techniques such as FDM, FDTD, MoM, and FEM have been discussed in Chapter 7 for analyzing more complex structures.

DIFFICULTIES OF COMPUTATIONAL MODELING

Every measured and computed value is uncertain, according to the precision of the measuring instrument and the number of significant digits carried along or according to the number of terms retained in the summation of a series. Consequently, all numerical solutions are approximate.

For the solution of numerical methods, one usually starts with an approximated mathematical model of a system with some initial data and then computes, after some intermediate steps, finally the result is obtained. So the error in the computed results may arise due to the following reasons:

(i)   mathematical modeling of the physical system

(ii) using of a digital computer for computing the numerical methods

(iii) uncertainty in physical experimental data

(iv) Machine error

(v) Mathematical truncation error.

NUMERICAL ERROR ANALYSIS

The term numerical error does not refer to a mistake. Rather, it refers to the ideas of deviation or uncertainty. The errors due to the above reasons can be classified into two types: (i) truncation error, and (ii) round off error. The error arises from the approximations to represent exact mathematical procedures, operations and quantities, which are called the truncation error. For example, errors due to finite representation of the infinite series expansion of the function $\sin x$, $\cos x$ and e^x *etc.,* are truncation errors.

On the other hand, the error created either due to the given data being approximate, or due to the limitation of the computing aids: mathematical tables, desk calculators, or the digital computer is called round off error. For example, numbers $\pi, e, \sqrt{7}$, and $\sqrt{101}$ cannot be expressed by a fixed number of significant figures during calculations using a digital computer. As the digital computer uses base-2 (binary) representation, so it cannot precisely represent certain exact base-10 numbers. The discrepancy introduced by this omission of significant figures is called round-off error.

However, for both types, the relationship between the exact, or true, result and the approximation can be formulated as:

$$\text{True error } (\in) = \text{True value } (x_T) - \text{Approximate value } (x_A) \qquad (1.3)$$

For many purposes, we prefer to study the percentage of relative error. The relative error can be defined by

$$\text{Relative error, } (\in_R) = \frac{\text{True error } (\in)}{\text{True value } (x_T)} = \frac{x_T - x_A}{x_T} \qquad (1.4)$$

The relative error can also be expressed in 100 percentages as follows:

$$\in_R = \frac{x_T - x_A}{x_T} \times 100\% \qquad (1.5)$$

In real world problems or in an actual situation, it is very difficult to obtain the true value. For such cases, an alternate, approximate error is used to calculate the error. Approximate error is defined as the difference between the present approximation and the previous approximation.

Approximate error (E_A)

$$= \text{present approximation} - \text{previous approximation} \qquad (1.6)$$

The relative approximate error $(\in_A)$ is defined as the ratio between the approximate error and the present approximation.

$$\textit{i.e.} \quad \text{relative approximate error, } (\in_R) = \frac{\text{Approximat e error}}{\text{present approximat ion}} \qquad (1.7)$$

The signs of (1.5) and (1.8) may be either positive or negative. However, during the computations, we may not be concerned with the sign of the error, but we are interested in whether the percent absolute value is lower than a pre-specified percent tolerance. For such a case, no further iterations are necessary and the process is stopped if

$$\left| \in_A \right| < \in_S \qquad (1.8)$$

where $\in_S$ is a pre-specified tolerance. If at least m significant digits are required to be correct in the final answer, then

$$|\in_A| \leq 0.5 \times 10^{2-m} \% \qquad (1.9)$$

Thus, numerical error analysis helps to determine the accuracy of numerical results. It also helps to develop stopping criteria for iterative algorithms.

Example 1.4 An approximate values of $x_A = e = 2.7142857 \ldots$ and its true value is $x_T = e = 2.7182818 \ldots$. Find the true error and relative error.

we know,

$$\text{True error, } \in = x_T - x_A = 2.7182818 .. - 2.7142857 .. = 0.003996 ..$$

$$\text{Relative error, } (\in_R) = \frac{x_T - x_A}{x_T} = \frac{0.003996 ..}{2.7182818 ..} = 0.00147 ..$$

Example 1.5 Suppose you have a task of measuring resistance of a resistor of an electrical system and it comes up 499.5 Ω. If the true value is 500 Ω, find the true error and relative error.

we know,

$$\text{True error, } \in = x_T - x_A = 500 - 499.5 = 0.5.$$

and

$$\text{Relative error, } (\in_R) = \frac{x_T - x_A}{x_T} = \frac{0.5}{500} = 0.001$$

Example 1.6 A exponential function $f(x) = 7e^{0.5x}$ is given at $x = 2$, find the (i) $f'(2)$ using $h = 0.3$, (ii) $f'(2)$ using $h = 0.15$, (iii) approximate error, and (iv) absolute relative approximate error.

Solution: (i) For $x = 2$ and $h = 0.3$

$$f'(x) \approx \frac{f(x+h)-f(x)}{h} \approx \frac{f(2+0.3)-f(2)}{0.3} \approx \frac{f(2.3)-f(2)}{0.3}$$

$$= \frac{7e^{0.5(2.3)}-7e^{0.5(2)}}{0.3} = 10.263$$

(i) For $x = 2$ and $h = 0.15$

$$f'(x) \approx \frac{f(2+0.15)-f(2)}{0.15} \approx \frac{f(2.15)-f(2)}{0.15}$$

$$= \frac{7e^{0.5(2.15)}-7e^{0.5(2)}}{0.15} = 9.8800$$

(iii) Approximate error (E_A)

= present approximation − previous approximation

= $9.8800 - 10.263 = -0.38300$.

(iv) Absolute relative approximate error

$$\left| \epsilon_R \right| = \left| \frac{\text{Approximat e error}}{\text{present approximat ion}} \right| = \left| \frac{-0.38300}{9.8800} \right| = 0.03865$$

ALGORITHM, COMPUTER LANGUAGE AND SOFTWARE

For many mathematical modelings, obtaining the solution, is very laborious and time consuming by hand. With the aid of digital computer, such calculations can be performed easily. However, once the mathematical modeling with method has been decided, we must describe a complete and unambiguous set of computational steps to be followed in a particular sequence to obtain the solution. This description is called an algorithm. The computer is concerned with algorithm and not with the method. An algorithm has five important features.

i. Finiteness: an algorithm must terminate after a finite number of steps

ii. Definiteness: each step of an algorithm must be clearly defined or the action to be taken must be unambiguously specified.

iii. Inputs: an algorithm must specify the quantities which must be read before the algorithm can begin.

iv. Outputs: an algorithm must specify the quantities which are to be outputted and their proper place.

v. Effectiveness: an algorithm must be effective, which means that all operations are executable.

The algorithm tells the computer where to start, what information to use, what operations to be carried out and in which order, what information to be printed and when to stop. Graphical representation of a specific sequence of steps (algorithm) is called a flow chart, which is followed by the computer to produce the solution of a given problem. The flow chart can be easily translated into any high level language, for example, FORTRAN, PASCAL, ALGOL, BASIC, *etc.* and can be executed on the computer.

In the past, FORTRAN, BASIC languages were used for scientific computation. Later, in the 1990s, C, C++ and Java were used too much for computation. Due to the increased power of a personal computer, MATLAB, MathCAD, Mapple and Mathematica are used for scientific computation. Among them, MATLAB has a tremendous influence to the computation of numerical methods. It is impossible to ignore the dominant place taken in recent years by MATLAB, which has been very successful in becoming the scientific programming language of choice for many applications.

Most examples are given in a generic pseudo-code which is heavily based on MATLAB, but occasionally, raw MATLAB code will be given. The author is still of the opinion that students should be comfortable in as many different languages as possible—even though most scientific programming today might be done in MATLAB, it is still the case that there is a lot of computer code (called legacy code) that is still being used and that was written in FORTRAN or Pascal or Algol or C.

There exist several sources for good mathematical software, which is given below:

International Mathematical and Statistical Library (IMSL): It is suitable for solving a large number of mathematical and statistical problems.

Numerical Algorithms Group (NAG):   This is also suitable for mathematical and statistical computation, but it is available in any one of three languages-FORTRAN, Pascal, Algol or C.

Linear Algebra Package (LINPACK): It contains FORTRAN subprograms for direct methods for general, symmetric, triangular and tridiagonal matrices.

More specialized packages have also been developed, notably QUADPACK (numerical integration), ITPACK (Iterative Methods), DEPACK (Differential equation solver), ELLPACK (Elliptic partial differential equations solver), EISPACK (eigenvalue methods), LAPACK (an updated package that combines and actually replaces LINPACK and EISPACK), and others.

CONSENT FOR PUBLICATION

Not applicable.

CONFLICT OF INTEREST

The author declares no conflict of interest, financial or otherwise.

ACKNOWLEDGEMENT

Declared none.

REFERENCES

[1] D. D. Moursund, C. S. Duris, *Elementary theory and applications of numerical analysis*, McGraw-Hill, New York, 1967.
[2] S. C. Chapra, R. P. Canale, *Numerical methods for engineers*, McGraw-Hill Education, New York, USA.
[3] M. K. Jain, S. R. K. Iyengar, and R. K. Jain, *Numerical methods for scientific and engineering computation*, Wiley Eastern Limited, India
[4] S. S. Sastry, *Introductory methods of numerical analysis*, Prentice-Hall, India, 2003.

EXERCISES

1. The measuring length of a bridge comes up 15999 cm. If the true value is 16000 cm, determine (i) the true error, and (ii) the relative error.
2. If $x = 3y^7 - 6y$, find the percentage error in x at $y = 1$, if the error in y is 0.05.
3. An approximate value of π is given by 3.1428571 and its true value is 3.1415926. Find the absolute and relative errors.

CHAPTER 2

Computational Solution of Nonlinear Equations

Md. Masud Rana[1*], Wei Xu[2] and Youguang Guo[3]

[1] Department of Electrical & Electronic Engineering, Rajshahi University of Engineering & Technology, Bangladesh

[2] Department of Electrical & Electronic Engineering, Huazhong University of Science and Technology (HUST), China

[3] Department of Electrical & Electronic Engineering, University of Technology Sydney (UTS), Australia

Abstract: In this chapter, computational methods for the solution of nonlinear equations, particularly the solution of transcendental equations, have been presented. Various types of computational methods are discussed with the engineering problem analysis.

Keywords: Algorithm, Transcendental, Non-linear, Iterative, Engineering, application.

INTRODUCTION

In the field of engineering and science, it is often required to find the roots of nonlinear equations [1]. Usually, the nonlinear equations are expressed in the form of a quadratic, higher-order polynomial or the form of transcendental functions, e.g. $1 + \sin x - 4x$, $x \tan x - \sinh x$, $e^x + \cos x$ *etc.* Though closed-form solutions are available to find the roots of a quadratic expression, to find the roots of other expressions is quite difficult. A lot of tasks, as well as time, are needed to find their roots. There are generally two types of methods, (i) direct method and (ii) iterative methods, which are used to solve the nonlinear equation with the form of a higher-order polynomial or transcendental or a combination of both [2, 3]. In this chapter, we describe iterative methods such as (i) bisection, (ii) false position, (iii) secant and (iv) newton Rapshon methods for the solution of a single nonlinear equation.

Corresponding author Md. Masud Rana: Department of Department of Electrical & Electronic Engineering, Rajshahi University of Engineering & Technology, Rajshahi, Bangladesh; E-mail: md.masud.rana.ruet@gmail.com

For the solution of polynomial equations, Miller and Lagurae methods are discussed, and finally, Newton is presented for the solution of nonlinear equations.

BISECTION METHOD

The bisection method is one of the simplest and most reliable iterative methods for the solution of nonlinear equations. This method is based on the repeated application of the intermediate value theorem. This theorem states that if   is real and continuous in the interval $a < x < b$ and   and   are   of   opposite signs,   that   is   then   there   is   at   least   one   real   root   in   the   interval between   and   . This method is also known as binary chopping of half chopping or half-interval method [4]. The method is shown graphically in Fig. (**2.1**).

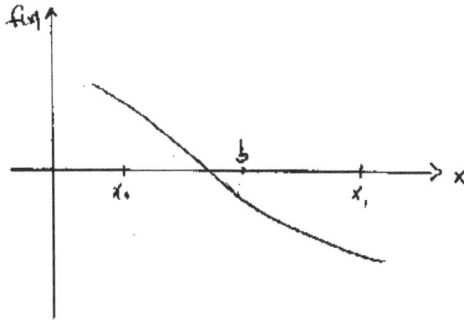

Fig. (2.1). Illustrating bisection method: at least one root exists between two points on the function that is real, continuous.

The solution procedure of the bisection method is discussed below.

i)   Choose two guess values a and b for finding the root of the function $f(x)$

   such that $f(a) * f(b) < 0$ or $f(x)$ changes sign between a and b.

ii)   Estimate the root x_m of the function $f(x)$ as the midpoint between a and b as

$$x_m = \frac{a+b}{2}.$$

iii) If $f(a) * f(b) = 0$; then the root is x_m. Otherwise, the root lies either between a and x_m or between x_m and b. If $f(a) * f(x_m) < 0$, the root lies between a and

x_m, hence $a = a$ and $b = x_m$. If $f(a) * f(x_m) > 0$, the root lies between x_m and b , hence $a = x_m$ and $b = b$.

iv) Find the new estimate of the root,   .

Pseudo Code of Bisection Method

Procedure Bisection $(n, f(x), a, b, \varepsilon)$

integer i

if

for $i = 1$ to n do

$$x_{mi} \leftarrow \frac{a_i + b_i}{2}$$

if

$a_{i+1} \leftarrow a_i$

$b_{i+1} \leftarrow x_{im}$

ifelse

$a_{i+1} \leftarrow x_{mi}$

$b_{i+1} \leftarrow b_i$

end if

end for

if $\left| \dfrac{x_{mi+1} - x_{mi}}{x_{mi+1}} \right| < \varepsilon$

endif
end procedure Bisection

Example 2.1 Use bisection method to find the smallest point root of the following equation:

$$f(x) = x^3 - 2x - 5 = 0$$

Solution

Since $f(2) < 0$ and $f(3) > 0$, a root lies between 2 and 3, therefore, $x_m = \dfrac{2+3}{2} = 2.5$. $f(x_m) = 5.6250$ and $f(2) * f(x_m) < 0$. Thus, the root lies in the interval $(2, 2.5)$. The sequence of the intervals are obtained as follows:

x_0	x_1	$f(x_0)$	$f(x_1)$	x_n	$f(x_n)$	Error
2	3	-1	16	2.5000	5.6250	11.1111
2	2.5000	-1	1.8906	2.2500	1.8906	5.8824
2	2.2500	-1	0.3457	2.1250	0.3457	3.0303
2	2.1250	-0.3513	0.3457	2.0625	-0.3513	1.4925
2.0625	2.1250	-0.0089	0.3457	2.0998	-0.0089	0.7407
2.0938	2.1250	-0.0089	0.1668	2.1094	0.1668	0.3717
2.0938	2.1094	-0.0089	0.0786	2.1016	0.0786	0.1862
2.0938	2.1016	-0.0089	0.0347	2.0977	0.0347	0.0932
2.0938	2.0977	-0.0089	0.0129	2.0957	0.0129	0.0466
2.0938	2.0957	-0.0089	0.0020	2.0947	0.0016	0.0233
2.0938	2.0947	-0.0089	0.0020	2.0947	-0.0035	0.0117
2.0942	2.0947	-0.0039	0.0020	2.0945	-0.0005	0.002

Root = 2.0945.

It is required to count the iterations and place a limit on the number of iterations that will be performed. Otherwise, the program could be trapped in an infinite loop. Also, it is better to test for the cases and $f(b) * f(x_m) > 0$. It may be that the function does not cross the x-axis between f_0 and f_1 or crosses more than once.

However, the bisection method may not be efficient because it does not take into consideration that $f(a)$ is much closer to the zero of the function $f(x)$ as compared to $f(b)$. In other words, the next predicted root x_m would be closer to a (in the example as shown in Fig. (**2.1**) than the mid-point between a and b. The false-position method takes advantage of this observation mathematically by drawing a secant from the function value at a and function value b and estimates the root where it crosses the x-axis.

FALSE POSITION METHOD

Based on two similar triangles, shown in Fig. (**2.2**), one gets:

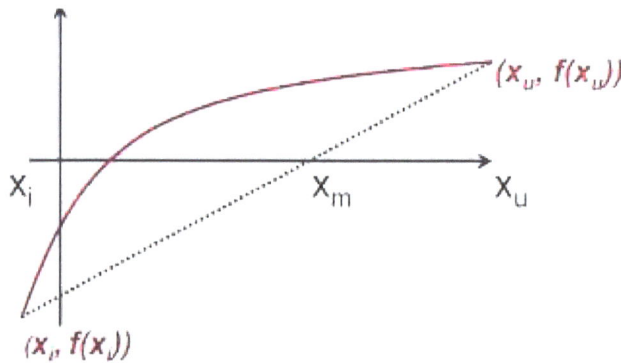

$$\frac{0 - f(x_L)}{x_r - x_L} = \frac{0 - f(x_U)}{x_r - x_U} \tag{2.1}$$

From Equation 2.1, we obtain,

$$(x_r - x_L)f(x_U) = (x_r - x_U)f(x_L) \tag{2.2}$$

$$x_U f(x_L) - x_L f(x_U) = x_r \{f(x_L) - f(x_U)\} \tag{2.3}$$

The above equation can be solved to obtain the next predicted root x_m as

$$x_r = \frac{x_U f(x_L) - x_L f(x_U)}{f(x_L) - f(x_U)} \tag{2.4}$$

The above equation, through simple algebraic manipulations, can also be expressed as

$$x_r = x_U - \frac{f(x_U)}{\left\{\dfrac{f(x_L)-f(x_U)}{x_L - x_U}\right\}} \tag{2.5}$$

Or,

$$x_r = x_L - \frac{f(x_L)}{\left\{\dfrac{f(x_U)-f(x_L)}{x_U - x_L}\right\}} \tag{2.6}$$

Observe the resemblance of Equations 2.5 and 2.6 by the secant method.

False-Position Algorithm

The steps to apply the false-position method to find the root of the equation $f(x)=0$ are as follows:

1. Choose x_L and x_U as two guesses for the root such that $f(x_L)f(x_U)<0$, or in other words, $f(x)$ changes sign between x_L and x_U.

2. Estimate the root, x_r of the equation $f(x)=0$ as,

$$x_r = \frac{x_U f(x_L) - x_L f(x_U)}{f(x_L)-f(x_U)}$$

3. Now check the following:

If $f(x_L)f(x_r)<0$, then the root lies between x_L and x_r; then $x_L = x_L$ and $x_U = x_r$

If $f(x_L)f(x_r)>0$, then the root lies between x_r and x_U; then $x_L = x_r$ and $x_U = x_U$

If $f(x_L)f(x_r)=0$, then the root is x_r. Stop the algorithm.

4. Find the new estimate of the root,

$$x_r = \frac{x_U f(x_L) - x_L f(x_U)}{f(x_L)-f(x_U)}$$

Find the absolute relative approximate error as,

$$|\epsilon_a| = \left| \frac{x_r^{new} - x_r^{old}}{x_r^{new}} \right| \times 100$$

Where,

x_r^{new} = estimated root from the present iteration
x_r^{old} = estimated root from the previous iteration

5. Compare the absolute relative approximate error $|\epsilon_a|$ with the pre-specified relative error tolerance ϵ_s. If $|\epsilon_a| > \epsilon_s$, then go to step 3, else stop the algorithm. Note that one should also check whether the number of iterations is more than the maximum number of iterations allowed. If so, one needs to terminate the algorithm and notify the user about it.

Note that the false-position and bisection algorithms are quite similar. The only difference is the formula used to calculate the new estimate of the root x_r, as shown in steps #2 and #4.

Example 2.2. Using the false position method, find the smallest point root of the following equation:

$$f(x) = x^3 - 6x^2 + 11x - 1 = 0$$

Solution:

Since $f(0.5) < 0$ and $f(1.75) > 0$, a root lies between 0.5 and 1.75 and therefore, $x_r = \dfrac{0.5 + 1.75}{2} = 1.6111 \cdot$ $f(x_r) = 5.3309$ and $f(a) * f(x_r) < 0$. Thus a root lies in the interval $(0.5, 1.6111)$. Similarly, the other two roots exist among the points $(1.25, 2.5)$ and $(2.80, 3.25)$. The details are explained using a Matlab code as follows:

%%%

(i) The Lower Value, x1=0.5

The Upper Value, x2=1.75

The Pre-specified Tolerance, Es=0.0001

iter	a	b	xr	f1	f2	Ea
1.0000	0.5000	1.7500	1.6111	-1.8750	0.3301	0
2.0000	0.5000	1.6111	1.4448	-1.8750	0.3841	0.1151
3.0000	0.5000	1.4448	1.2842	-1.8750	0.3490	0.1251
4.0000	0.5000	1.2842	1.1611	-1.8750	0.2485	0.1060
5.0000	0.5000	1.1611	1.0837	-1.8750	0.1470	0.0714
6.0000	0.5000	1.0837	1.0413	-1.8750	0.0775	0.0408
7.0000	0.5000	1.0413	1.0198	-1.8750	0.0384	0.0211
8.0000	0.5000	1.0198	1.0094	-1.8750	0.0185	0.0103

9.0000	0.5000	1.0094	1.0044	-1.8750	0.0087	0.0049
10.0000	0.5000	1.0044	1.0021	-1.8750	0.0041	0.0023
11.0000	0.5000	1.0021	1.0010	-1.8750	0.0019	0.0011
12.0000	0.5000	1.0010	1.0004	-1.8750	0.0009	0.0005
13.0000	0.5000	1.0004	1.0002	-1.8750	0.0004	0.0002
14.0000	0.5000	1.0002	1.0001	-1.8750	0.0002	0.0001
15.0000	0.5000	1.0001	1.0000	-1.8750	0.0001	0.0001

Root = 1.0000

(ii) The Lower Value, x1=1.25

The Upper Value, x2=2.50

The Pre-specified Tolerance, Es=0.0001

iter	a	b	xr	f1	f2	Ea

1.0000	1.2500	2.5000	1.8333	0.1620	-0.3750	0
2.0000	1.8333	2.5000	2.0345	0.1620	-0.0344	0.0989
3.0000	1.8333	2.0345	1.9992	0.0008	-0.0344	0.0176
4.0000	1.9992	2.0345	2.0000	0.0008	-0.0000	0.0004
5.0000	1.9992	2.0000	2.0000	0.0000	-0.0000	0.0000

Root = 2.0000

(iii) The Lower Value, x1=2.80

The Upper Value, x2=3.25

The Pre-specified Tolerance, Es=0.0001

iter	a	b	xr	f1	f2	Ea
1.0000	2.8000	3.2500	2.9308	-0.1244	0.7031	0
2.0000	2.9308	3.2500	2.9788	-0.0411	0.7031	0.0161
3.0000	2.9788	3.2500	2.9938	-0.0124	0.7031	0.0050
4.0000	2.9938	3.2500	2.9982	-0.0036	0.7031	0.0015
5.0000	2.9982	3.2500	2.9995	-0.0010	0.7031	0.0004
6.0000	2.9995	3.2500	2.9998	-0.0003	0.7031	0.0001
7.0000	2.9998	3.2500	3.0000	-0.0001	0.7031	0.0000

Root = 3.0000

The convergent root of the given equation is = 1, 2 & 3.

RATE OF CONVERGENCE

If the function $f(x)$ in the equation $f(x) = 0$ is convex in the interval (x_0, x_1) that contains a root, then one of the points x_0 or x_1 is always fixed, and the other point varies with k. If the point x_0 is fixed, then the function $f(x)$ is approximated by the straight line passing through the points (x_0, f_0) and $(x_k, f_k), k = 1, 2 .....$

we assume that   is a simple root of $f(x) = 0$. Substituting $x_k = \xi + \varepsilon_k$ in

$$x_{k+1} = x_k - \frac{f(x_k)(x_k - x_{k-1})}{f(x_k) - f(x_{k-1})} \text{ and then we obtain,}$$

$$\varepsilon_{k+1} = \varepsilon_k - \frac{f(\xi + \varepsilon_k)(\varepsilon_k - \varepsilon_{k-1})}{f(\xi + \varepsilon_k) - f(\xi + \varepsilon_{k-1})} \tag{2.7}$$

Expanding $f(\xi + \varepsilon_k)$ and $f(\xi + \varepsilon_{k-1})$ in Taylor's series about the point   and by considering $f(\xi) = 0$ we get,

$$\varepsilon_{k+1} = \varepsilon_k - \frac{(\varepsilon_k - \varepsilon_{k-1})\left[\varepsilon_k f'(\xi) + \frac{1}{2}\varepsilon_k^2 f''(\xi) + \ldots\ldots\right]}{(\varepsilon_k - \varepsilon_{k-1})f'(\xi) + \frac{1}{2}\left(\varepsilon_k^2 - \varepsilon_{k-1}^2\right)f''(\xi) + \ldots\ldots} \tag{2.8}$$

$$\varepsilon_{k+1} = \varepsilon_k - \left[\varepsilon_k + \frac{1}{2}\varepsilon_k^2 \frac{f''(\xi)}{f'(\xi)} + \ldots\right]\left[1 + \frac{1}{2}(\varepsilon_k + \varepsilon_{k-1})\frac{f''(\xi)}{f'(\xi)} + \ldots\right]^{-1} \tag{2.9}$$

or

$$\varepsilon_{k+1} = \frac{1}{2}\varepsilon_k \varepsilon_{k-1}\frac{f''(\xi)}{f'(\xi)} + O\left(\varepsilon_k^2 \varepsilon_{k-1} + \varepsilon_{k-1}^2 \varepsilon_k\right) \tag{2.10}$$

$\varepsilon_{k+1} = C\varepsilon_k\varepsilon_{k-1}$ where $C = \frac{1}{2}\frac{f''(\xi)}{f'(\xi)}$ and the higher power of ε_k are neglected.

According to the Regula-Falsi method, Equ. 2.10 becomes:

$$\varepsilon_{k+1} = C\varepsilon_k\varepsilon_0 \tag{2.11}$$

Where, $C = \frac{f''(\xi)}{2f'(x_n)}$ and $\varepsilon_0 = x_0 - \xi$ is independent of . Therefore, we can write, $\varepsilon_{k+1} = C'\varepsilon_k$, where $C' = C\varepsilon_0$ is the asymptotic error constant. Hence, the Regula-Falsi method has a linear rate of convergence.

NEWTON RAPHSON METHOD

Finding the roots of a nonlinear equation using bisection and false position methods requires bracketing of the root by two guesses. Such methods are called bracketing methods. These methods are always convergent as they are based on reducing the interval between the two guesses so as to obtain zero in the root of the equation.

On the other hand, the root in the Newton-Raphson method is not bracketed. In fact, only one initial guess of the root is needed to get the iterative process for finding the root of an equation. So, the method falls in the category of open methods. Convergence in open methods is not guaranteed, but if the method does converge, it is much faster than the bracketing methods.

Derivation

The Newton-Raphson method is based on the principle that if the initial guess of the root of $f(x) = 0$ is at x_i, then if one draws the tangent to the curve at $f(x_i)$, the point$_{i+1}$, where the tangent crosses the x-axis, is an improved estimate of the root (Fig. **2.3**). Using the definition of the slope of a function, at $x = x_i$

$$f'(x_i) = \tan\theta = \frac{f(x_i) - 0}{x_i - x_{i+1}}$$

(2.12)

which gives

$$x_{i+1} = x_i - \frac{f(x_i)}{f'(x_i)}$$

(2.13)

Eq. 2.12 is called the Newton-Raphson formula for solving nonlinear equations of the form $f(x) = 0$. So, starting with an initial guess, x , one can find the next guess$_{i+1}$, by using Eq. 2.12. One can repeat this process until one finds the root within a desirable tolerance. The method is shown graphically in Fig. (**2.3**).

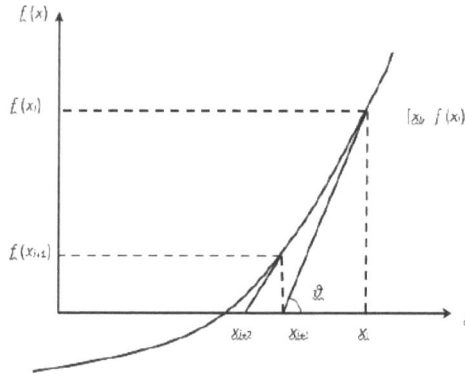

Fig. (2.3). Geometrical illustration of the Newton-Raphson method.

Algorithm

The steps of the Newton-Raphson method to find the root of an equation are

1. Evaluate $f'(x)$ symbolically

2. Use an initial guess of the root, to estimate the new value of the root, x_{i+1}, as

$$x_{i+1} = x_i - \frac{f(x_i)}{f'(x_i)}$$

3. Find the absolute relative approximate error $|\in_a|$ as

$$|\in_a| = \left| \frac{x_{i+1} - x_i}{x_{i+1}} \right| \times 100$$

4. Compare the absolute relative approximate error with the pre-specified relative error tolerance, $\in_s$. If $|\in_a| > \in_s$, then go to Step 2, else stop the algorithm. Also, check if the number of iterations has exceeded the maximum number of iterations allowed. If so, one needs to terminate the algorithm and notify the user.

Example 2.3. Using the Newton Raphson method, find the smallest point root of the following equation:

$$f(x) = x\sin x + \cos x = 0$$

Solution:

We have $f(x) = x\sin x + \cos x = 0$, and $f'(x) = x\cos x$. The iteration formula is, therefore, $x_{r+1} = x_r - \dfrac{x_r \sin x_r + \cos x_r}{x_r \cos x_r}$. With $x_0 = 3.1416$, the successive iterations are as follows:

iter.	x_r	$f(x_r)$	x_{r+1}	Error
1	3.1416	-0.0662	2.8233	1.0000
2	2.8233	-0.0006	2.7986	0.0008
3	2.7986	-0.0000	2.7984	0.0001

The required root is 2.7984 and correct upto 4 decimal places.

CONVERGENCE OF NEWTON RAPHSON METHOD

According to Taylor's theorem, any function $f(x)$, which has a continuous second derivative, can be represented by an expansion about a point that is close to a root of $f(x)$. Suppose x_r is the root of $f(x)$ and x_n is an estimate of x_r, i.e., $|x_r - x_n| = \delta \ll 1$. Then, the expansion of $f(x_r)$ regarding x_n is:

$$0 = f(x_r) = f(x_n + \delta) = f(x_n) + f'(x_n)(x_r - x_n) + \frac{f''(\xi)}{2}(x_r - x_n)^2 \qquad (2.14)$$

for some ξ between x_n and x_r. By Newton Raphson method, we know that

$$x_{n+1} = x_n - \frac{f(x_n)}{f'(x_n)} \qquad (2.15)$$

$$i.e., \ f(x_n) = f'(x_n)(x_n - x_{n+1}) \qquad (2.16)$$

Using Eq. 2.16 in Eq. 2.14, we get

$$0 = f'(x_n)(x_r - x_{n+1}) + \frac{f''(\xi)}{2}(x_r - x_n)^2 \tag{2.17}$$

$e_n = (x_r - x_n)$, $e_{n+1} = (x_r - x_{n+1})$, where e_n, e_n denote the error in the solution at n^{th} and $(n+1)^{th}$ iterations

$$e_{n+1} = -\frac{f''(\xi)}{2f'(x_n)} \cdot e_n^2 \tag{2.18}$$

$$\text{or, } e_{n+1} \propto e_n^2 \tag{2.19}$$

Eq. 2.19 shows that the rate of convergence is at least quadratic if the following conditions are satisfied.

1. $f'(x) \neq 0$; for all $x \in I$, where I is the interval $[x_r - r, x_r + r]$ for some $r \geq |x_r - x_0|$.

2. $f''(x)$ is continuous, for all $x \in I$

3. x_0 is sufficiently close to the root

The term in this context means the following:

a. Taylor approximation is accurate enough such that we can ignore higher-order terms.

b. $\frac{1}{2}\left|\frac{f''(x_n)}{f'(x_n)}\right| < C\left|\frac{f''(x_r)}{f'(x_r)}\right|$, for some $C < \infty$

c. $C\left|\frac{f''(x_r)}{f'(x_r)}\right|e_n < 1$, for $n \in Z$, $n \geq 0$ and C satisfying condition

Finally, Eq. 2.19 can be expressed in the following way:

$|e_{n+1}| \le M e_n^2$; where M is the supremum of the variable coefficient of e_n^2 on the interval I defined in condition 1:

$$M = \sup_{x \in I} \frac{1}{2} \left| \frac{f''(x_n)}{f'(x_n)} \right| \tag{2.20}$$

The initial point x_0 has to be chosen such that conditions 1 to 3 are satisfied, where the third condition requires $M|e_0| < 1$.

SECANT METHOD

The Newton-Raphson method of solving a nonlinear equation $f(x) = 0$ is given by the iterative formula

$$x_{i+1} = x_i - \frac{f(x_i)}{f'(x_i)} \tag{2.21}$$

One of the drawbacks of the Newton-Raphson method is the evaluation of the derivative of the function. With the availability of symbolic manipulators such as Maple, MathCAD, MATHEMATICA and MATLAB, this process has become more convenient. However, it still can be a laborious process and even intractable if the function is derived as part of a numerical scheme. To overcome these drawbacks, the derivative of the function, is approximated as

$$f'(x_i) = \frac{f(x_i) - f(x_{i-1})}{x_i - x_{i-1}} \tag{2.22}$$

Substituting Eq. 2.22 in Eq. 2.21 gives:

$$x_{i+1} = x_i - \frac{f(x_i)(x_i - x_{i-1})}{f(x_i) - f(x_{i-1})} \tag{2.23}$$

The above equation is called the secant method equation. This method requires two initial guesses, but unlike the bisection method, the two initial guesses do not need to bracket the root of the equation. The secant method is an open method and may or may not converge. However, when the secant method converges, it typically faster than the bisection method. However, since the derivative is approximated, as given by Eq. 2.23, it typically converges slower than the Newton-Raphson method.

The secant method can also be derived from geometry, as shown in Fig. **(2.4)**. Taking two initial guesses, x_{i-1} and x_i, one draws a straight line between and $f(x_{i-1})$ passing through the x-axis at x_{i+1}. *ABE* and *DCE* are similar triangles.

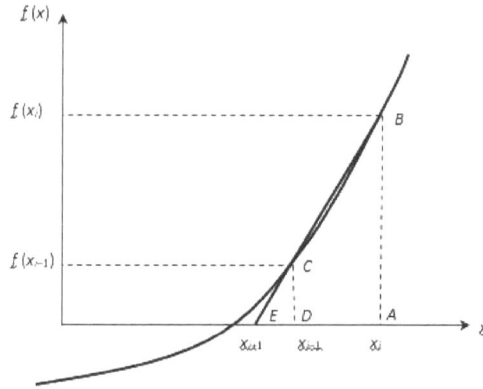

Fig. (2.4). Geometrical representation of the secant method.

Hence,

$$\frac{AB}{AE} = \frac{DC}{DE} \qquad (2.24)$$

$$i.e., \; \frac{f(x_i)}{x_i - x_{i+1}} = \frac{f(x_{i-1})}{x_{i-1} - x_{i+1}} \qquad (2.25)$$

On rearranging, the secant method is given as

$$x_{i+1} = x_i - \frac{f(x_i)(x_i - x_{i-1})}{f(x_i) - f(x_{i-1})} \qquad (2.26)$$

Algorithm 2.4. Secant Method

1. Read a, b

Remark: a, b are two initial guesses root.

2. f0 = f(a)

3. f1 = f(b)

4. *for* i = 1 to 100 in steps of 1 do

5. c=(a*f1-b*f0)/(f1-f0);

6. f2= f(c)

7. result(i,:)=[i,c,f2];

8. if |f2| < 0.0001 then GOTO 14

9. f0=f1

10. f1=f2

11. a = b

12. b = c

end for

13. Write ' Iteration   Xi-1   Xi   Xi+1   f(Xi-1)   f(Xi)'

14. Write result

15. Stop

Example 2.4. Using the Secant method, find the smallest point root of the following equation:

$$f(x) = e^{-x} - x = 0$$

Solution:

For finding the smallest positive root, we start with two trial points, $x_{i-1} = 0$ and

$x_i = 5.0000$. We now find the root using Algorithm 2.4 as follows:

Enter The Value of a = 0

Enter The Value of b = 5

Iteration	Xi-1	Xi	Xi+1	f(Xi-1)	f(Xi)
1.0000	0	5.0000	0.8343	1.0000	-4.9933
2.0000	5.0000	0.8343	0.4714	-4.9933	-0.4001
3.0000	0.8343	0.4714	0.5717	-0.4001	0.1527
4.0000	0.4714	0.5717	0.5672	0.1527	-0.0071
5.0000	0.5717	0.5672	0.5671	-0.0071	-0.0001

FINDING ROOTS OF POLYNOMIAL EQUATIONS

In the previous sections, we have investigated numerical methods for finding the roots of general functions. These techniques may sometimes work for finding the roots of a polynomial, but often problems arise because of multiple roots or complex roots. Various types of methods are available for nonlinear functions with these types of roots [2]. Among them, three methods, such as Muller and Laguerre's method, will be discussed here.

MULLER'S METHOD

Muller's method generalizes the secant method but uses quadratic interpolation among three points instead of linear interpolation between the two. Solving for the zeros of the quadratic allows the method to find complex pairs of roots. Given three previous guesses for the root x_{i-2}, x_{i-1}, x_i and the polynomial $P(x)$ at those points, the next approximation x_{i+1} is produced by the following formula:

$$P(x) = A(x - x_i)^2 + B(x - x_i) + y_i \tag{2.27a}$$

$$y_{i-1} = A(x_{i-1} - x_i)^2 + B(x_{i-1} - x_i) + y_i \tag{2.27b}$$

$$y_{i-2} = A(x_{i-2} - x_i)^2 + B(x_{i-2} - x_i) + y_i \tag{2.27c}$$

$$A = \frac{(x_{i-2} - x_i)(y_{i-1} - y_i) - (x_{i-1} - x_i)(y_{i-2} - y_i)}{(x_{i-1} - x_{i-2})(x_{i-1} - x_i)(x_{i-2} - x_i)} \tag{2.27d}$$

$$B = \frac{(x_{i-2} - x_i)^2 (y_{i-1} - y_i) - (x_{i-1} - x_i)^2 (y_{i-2} - y_i)}{(x_{i-1} - x_{i-2})(x_{i-1} - x_i)(x_{i-2} - x_i)}$$ (2.27e)

$$x_{i+1} - x_i = \frac{-B \pm \sqrt{B^2 - 4Ay_i}}{2A}$$ (2.28a)

$$x_{i+1} - x_i = -\frac{2y_i}{B \pm \sqrt{B^2 - 4Ay_i}}$$ (2.28b)

Where, the sign in the denominator is chosen to make its absolute value or modulus as large as possible. You can start the iterations with any three values of x that you like, e.g., three equally spaced values on the real axis. Note that you must allow for the possibility of a complex denominator, and subsequent complex arithmetic in implementing the method.

Muller's method is sometimes also used for finding complex zeros of analytic functions (not just polynomials) in the complex plane, for example, in the IMSL routine ZANLY [3].

LAGUERRE'S METHOD

Laguerre's method is by far the most straightforward of these general, complex methods. It does require complex arithmetic, even while converging to real roots; however, for polynomials with all real roots, it is guaranteed to converge to a root from any starting point. For polynomials with some complex roots, there is little theoretical proof about the method's convergence. Most of the empirical experience, suggests that non-convergence is extremely unusual and, can almost always be fixed by a simple scheme to break a nonconverging limit cycle.

In some instances, the complex arithmetic in the Laguerre method is no disadvantage since the polynomial itself may have complex coefficients. To motivate (although not rigorously derive) the Laguerre formulas, we can note the following relations between the polynomial and its roots and derivatives:

$$P(x) = (x - x_1)(x - x_2)\ldots\ldots(x - x_n)$$ (2.29)

$$\ln|P(x)| = \ln|x - x_1| + \ln|x - x_2| + \ldots\ldots + \ln|x - x_n|$$ (2.30)

$$\frac{d \ln |P(x)|}{dx} = \frac{1}{x-x_1} + \frac{1}{x-x_2} + \ldots\ldots + \frac{1}{x-x_n} = \frac{P'_n}{P_n} = G \tag{2.31}$$

$$\frac{d^2 \ln |P(x)|}{dx^2} = \frac{1}{(x-x_1)^2} + \frac{1}{(x-x_2)^2} + \ldots\ldots$$

$$\ldots\ldots + \frac{1}{(x-x_n)^2} = \left[\frac{P'_n}{P_n}\right]^2 - \frac{P''_n}{P_n} = H \tag{2.32}$$

The root x_1 that we seek is assumed to be located some distance a from our current guess x, while all other roots are assumed to be located at a distance b:

$$x - x_1 = a \quad ; \quad x - x_i = b \quad i = 2,3,\ldots.n \tag{2.33}$$

$$\frac{1}{a} + \frac{n-1}{b} = G \tag{2.34}$$

$$\frac{1}{a^2} + \frac{n-1}{b^2} = H \tag{2.35}$$

$$a = \frac{n}{G \pm \sqrt{(n-1)(nH-G^2)}} \tag{2.36}$$

Where, the sign should be taken to yield the largest magnitude for the denominator. Since the factor inside the square root can be negative, a can be complex.

FINDING THE ROOT OF THE SYSTEM OF NONLINEAR EQUATIONS

In the previous section, the focus of root finding is restricted only to a single equation. In this section, the process of root finding for a system of equations has been discussed. Consider a general system of n nonlinear equations in n unknowns:

$$f_i(x_1, x_2, \ldots\ldots, x_n) = 0, \quad \text{where } i = 1,2,3,\ldots\ldots, n. \tag{2.37}$$

The solution of n simultaneous, nonlinear equations is a much more formidable task than finding the root of a single equation. For solving the nonlinear simultaneous equations, there is no reliable method available in the literature. Therefore, we cannot provide the solution algorithm with a guaranteed good starting value of x, unless such a value is suggested by the physics of the problem.

The simplest and the most effective means of computing x is the Newton–Raphson method. It works well with simultaneous equations, provided that it is supplied with a good starting point. There are other methods that have better global convergence characteristics, but all are mere variants of the Newton–Raphson method. The Newton Raphson method for the root finding of the system of equations is discussed below.

Consider the following equation

$$f(x,y) = 0 \qquad\qquad (2.38)$$

$$g(x,y) = 0 \qquad\qquad (2.39)$$

Let the root be (x_0, y_0). Expanding $f(x,y)$ and $g(x,y)$ using the Taylor series near the roots (x_0, y_0), we obtain:

$$f(x_0,y_0) = f(x,y) + \left.\frac{\partial f}{\partial x}\right|_{x,y}(x_0 - x) + \left.\frac{\partial f}{\partial y}\right|_{x,y}(y_0 - y) + \text{HO terms} \qquad (2.40a)$$

$$g(x_0,y_0) = g(x,y) + \left.\frac{\partial g}{\partial x}\right|_{x,y}(x_0 - x) + \left.\frac{\partial g}{\partial y}\right|_{x,y}(y_0 - y) + \text{HO terms} \qquad (2.40b)$$

Omitting the higher-order terms and letting $(x_0 - x = \Delta x)$ and $(y_0 - y = \Delta y)$, we obtain (remember that $f(x_0,y_0) = 0$).

If the Jacobian matrix

$$J(f,g) = \begin{vmatrix} \dfrac{\partial f}{\partial x} & \dfrac{\partial f}{\partial y} \\[2mm] \dfrac{\partial g}{\partial x} & \dfrac{\partial g}{\partial y} \end{vmatrix} \qquad\qquad (2.41)$$

does not vanish, then the linear equation 2.40 possesses a unique solution given by

$$x_0 - x = \left. \begin{vmatrix} f & \dfrac{\partial f}{\partial y} \\[2mm] -g & \dfrac{\partial g}{\partial y} \end{vmatrix} \right/ J(f,g) \qquad\qquad (2.42)$$

$$y_0 - y = \left. \begin{vmatrix} \dfrac{\partial f}{\partial x} & -f \\ \dfrac{\partial g}{\partial x} & -g \end{vmatrix} \middle/ J(f,g) \right. \qquad (2.43)$$

The new approximation given by

$$x = x_0 - \left. \begin{vmatrix} f & \dfrac{\partial f}{\partial y} \\ -g & \dfrac{\partial g}{\partial y} \end{vmatrix} \middle/ J(f,g) \right. \qquad (2.44)$$

$$y = y_0 - \left. \begin{vmatrix} \dfrac{\partial f}{\partial x} & -f \\ \dfrac{\partial g}{\partial x} & -g \end{vmatrix} \middle/ J(f,g) \right. \qquad (2.45)$$

The process is to be repeated till we obtain the desired accurate roots. If the iteration converges, it does so quadratically.

CONSENT FOR PUBLICATION

Not applicable.

CONFLICT OF INTEREST

The author declares no conflict of interest, financial or otherwise.

ACKNOWLEDGEMENT

Declared none.

REFERENCES

[1]   D. D. Moursund, C. S. Duris, *Elementary theory and applications of numerical analysis*, McGraw-Hill, New York, 1967.
[2]   S. C. Chapra, R. P. Canale, *Numerical methods for engineers*, McGraw-Hill Education, New York, USA.
[3]   M. K. Jain, S. R. K. Iyengar, and R. K. Jain, *Numerical methods for scientific and engineering computation*, Wiley Eastern Limited, India.
[4]   S. S. Sastry, *Introductory methods of numerical analysis*, Prentice-Hall, India, 2003.

EXERCISES

1. For a 10K3A Betatherm thermistor, the relationship between the resistance R of the thermistor and the temperature is given by:

$$\frac{1}{T} = 1.129241 \times 10^{-3} + 2.341077 \times 10^{-4} \ln(R) + 8.775468 \times 10^{-8} \{\ln(R)\}^3$$

   where T is in Kelvin and R is in ohms. A thermistor error of no more than $\pm 0.01\,°C$ is acceptable. Use false position and bisection methods of finding the range of the resistance that is within this acceptable limit at $19\,°C$.

2. To find the inverse of a number a, one can use the equation

$$f(c) = a - \frac{1}{c} = 0$$

   where c is the inverse of a. Use the secant method of finding roots of equations to find the inverse of $a = 2.5$. Conduct three iterations to estimate the root of the above equation.

3. The speed v of a Saturn V rocket in vertical flight near the surface of the earth can be approximated by:

$$v = u \ln \frac{M_0}{M_0 - mt} - gt$$

   where, u is the velocity of the exhaust relative to the rocket, M_0 the mass of the rocket at liftoff, m the rate of fuel consumption, g the gravitational acceleration, and t the time measured from liftoff. Use the bisection and secant methods of finding roots and conduct three iterations to estimate the root.

4. The aluminum W310×202 (wide flange) column is subjected to an eccentric axial load P. The maximum compressive stress in the column is given by:

$$\sigma_{max} = \sigma \left[1 + \frac{ec}{r^2} \sec\left(\frac{L}{2r} \sqrt{\frac{\sigma}{E}} \right) \right]$$

   where $\bar{\sigma} = P/A$ =average stress, A is the cross-sectional area of the column, e the eccentricity of the load, c the half-depth of the column, and r the radius of gyration of the cross section.

5. An oscillating current in an electric circuit is described by $i = 9e^{-t} \sin(2\pi t)$, where t is in seconds. Determine the lowest value of Newton Raphson method using t such that $i = 3.6$.

CHAPTER 3

Computational Solution of Linear System Equations

Md. Masud Rana[1]*, Wei Xu[2] and Youguang Guo[3]

[1] *Department of Electrical & Electronic Engineering, Rajshahi University of Engineering & Technology, Bangladesh*

[2] *Department of Electrical & Electronic Engineering, Huazhong University of Science and Technology (HUST), China*

[3] *Department of Electrical & Electronic Engineering, University of Technology Sydney (UTS), Australia*

Abstract: In this chapter, computational methods for the solution of linear system equations have been presented. At first, the fundamentals of a matrix with matrix operation have been discussed. Then, various types of computational methods, such as Gauss elimination, Gauss-Jordan, Gauss Seidel, *etc.*, for the solution of linear system equations are discussed with the engineering problem analysis.

Keywords: Matrix, Gauss elimination, Linear, Algebraic, Simultaneous, engineering, Application.

INTRODUCTION

Recently Linear algebraic equations have been using in all branches of science and engineering. However, their most visible application is in engineering (particularly structures, elastic solids, heat flow, fluids, electromagnetic fields, electric circuits, *etc.*) for the analysis of linear systems. It is almost impossible to carry out the numerical analysis of the mentioned physical problems without encountering linear algebraic equations. However, for the discrete linear system, such as a truss or an electric circuit, its analysis leads directly to linear algebraic equations [1, 2]. In the case of a continuous system, which is described by differential equations or other operators, algebraic equations are obtained by approximating the differential equation with a system of algebraic equations. The well-known finite difference,

---
*Corresponding author **Md. Masud Rana:** Department of Electrical & Electronic Engineering, Rajshahi University of Engineering & Technology, Bangladesh; E-mail: md.masud.rana.ruet@gmail.com

finite element, and boundary element methods of analysis are used for approximation [3].

The modeling of linear systems invariably gives rise to equations of the form $[A][X]=[B]$, where $[B]$ is the input and $[X]$ represents the response of the system. The coefficient matrix $[A]$, which reflects the characteristics of the system, is independent of the input. Therefore, it is desirable to have an equation-solving algorithm that can handle any type of linear algebraic equation that arises from physical systems. We cannot possibly discuss all the special algorithms in the limited space available. The best we can do is present the basic methods of solution which are useful for engineers. Before discussing the matrix algorithm, we introduce some fundamental properties of a matrix.

FUNDAMENTALS OF A MATRIX

Consider the system of m linear equations in n unknowns:

$$a_{11}x_1 + a_{12}x_2 + \dots\dots + a_{1n}x_n = b_1$$
$$a_{21}x_1 + a_{22}x_2 + \dots\dots + a_{2n}x_n = b_2 \qquad \text{(3.1)}$$
$$\dots\dots\dots\dots\dots\dots\dots\dots\dots\dots\dots$$
$$a_{m1}x_1 + a_{m2}x_2 + \dots\dots + a_{mn}x_n = b_m$$

The above equations can be written as follows:

$$[A][X]=[B] \qquad \text{(3.2)}$$

where

$$A = \begin{bmatrix} a_{11} & a_{12} & \dots\dots & a_{1n} \\ a_{21} & a_{22} & \dots\dots & a_{2n} \\ \dots & \dots & \dots & \dots \\ a_{m1} & a_{m2} & \dots & a_{mn} \end{bmatrix} = [a_{ij}]$$

$$X = \begin{bmatrix} x_1 \\ x_2 \\ \vdots \\ x_n \end{bmatrix} \text{ and } B = \begin{bmatrix} b_1 \\ b_2 \\ \vdots \\ b_n \end{bmatrix}$$

where   matrix is an $m \times n$ array of numbers. Here,   is the number of rows and n is the number of columns. The coefficients a_{11}, a_{12} ..... are its elements and may be either real or complex. If $m = n$, the matrix is said to be a square matrix. If the matrix has only one column (row), it is said to be a column (row) matrix. If all the elements of the square matrix are zero ($a_{ij} = 0$), then the matrix is said to be a null matrix. The matrix is said to be diagonal if $a_{ij} = 0$, $i \neq j$. For example,

$$[A] = \begin{bmatrix} a_{11} & 0 & 0 \\ 0 & a_{22} & 0 \\ 0 & 0 & a_{33} \end{bmatrix} = [D] \tag{3.3}$$

The diagonal matrix is said to be a unit matrix if the elements of diagonal elements are equal to one. Matrix is said to be an upper triangular matrix if $a_{ij} = 0$, $i > j$, and a lower triangular matrix if $a_{ij} = 0$, $i < j$. For example,

$$[A] = \begin{bmatrix} a_{11} & a_{12} & a_{13} \\ 0 & a_{22} & a_{23} \\ 0 & 0 & a_{33} \end{bmatrix} = [U] \tag{3.4}$$

is an upper triangular matrix, and

$$[L] = \begin{bmatrix} a_{11} & 0 & 0 \\ a_{21} & a_{22} & 0 \\ a_{31} & a_{32} & a_{33} \end{bmatrix} = [L] \tag{3.5}$$

is a lower triangular matrix.

Matrix   is said to be symmetric if $a_{ij} = a_{ji}$; if $a_{ij} = -a_{ji}$, it is said to be a skew-symmetric. For example,

$$[A] = \begin{bmatrix} a_{11} & a_{12} & a_{13} \\ a_{12} & a_{22} & a_{23} \\ a_{13} & a_{23} & a_{33} \end{bmatrix} \tag{3.6}$$

is a symmetric matrix, and

$$[A] = \begin{bmatrix} a_{11} & a_{12} & a_{13} \\ -a_{12} & a_{22} & a_{23} \\ -a_{13} & -a_{23} & a_{33} \end{bmatrix} \tag{3.7}$$

is a skew-symmetric matrix.

Matrix is said to be a tri-diagonal matrix if $a_{ij} = 0$, for $|i - j| > 1$. For example,

$$[A] = \begin{bmatrix} a_{11} & a_{12} & 0 & 0 \\ a_{21} & a_{22} & a_{23} & 0 \\ 0 & a_{32} & a_{33} & a_{34} \\ 0 & 0 & a_{43} & a_{44} \end{bmatrix} \tag{3.8}$$

is a tri-diagonal matrix. Matrix is associated with a number called its determinant, which is as follows:

$$|A| = \begin{vmatrix} a_{11} & a_{12} & \cdots\cdots & a_{1n} \\ a_{21} & a_{22} & \cdots\cdots & a_{2n} \\ \cdots\cdots & \cdots\cdots & \cdots\cdots & \cdots\cdots \\ a_{m1} & a_{m2} & \cdots\cdots & a_{mn} \end{vmatrix} \tag{3.9}$$

if, then is said to be a nonsingular matrix; otherwise, it is said to be singular. Thus,

$$[A] = \begin{bmatrix} a_{11} & a_{12} & 0 \\ a_{12} & a_{22} & 0 \\ a_{13} & a_{23} & 0 \end{bmatrix} \tag{3.10}$$

is singular since $|A| = 0$.

MATRIX OPERATION

1. Equality of Two Matrices

Two matrices, say $[A]$ and $[B]$ are said to be equal if they are of the same size and if their corresponding elements are equal, that is, $[A]=[B]$ if $a_{ij}=b_{ij}$ for all i and j.

2. Addition & Subtraction

Two matrices $[A]$ and $[B]$ of the same size can be added by adding their corresponding elements. The elements of the resulting matrix $[C]$ are as follows:

$$[C]=[A]+[B] \tag{3.11}$$

i.e., $c_{ij}=a_{ij}+b_{ij}$

Similarly, the subtraction of two matrices, say, $[E]$ and $[F]$, is obtained by subtracting their corresponding elements as follows:

$$[D]=[E]-[F] \tag{3.12}$$

i.e., $d_{ij}=e_{ij}-f_{ij}$

Both matrix addition and subtraction are commutative:

$$[A]+[B]=[B]+[A] \tag{3.13}$$

Matrix addition and subtraction are also associative:

$$([A]+[B])+[C]=[A]+([B]+[C]) \tag{3.14}$$

3. Multiplication

Two matrices $[A]$ and $[B]$ can be multiplied only if the number of columns of $[A]$ is equal to the number of rows of $[B]$. Thus if $[A]$ and $[B]$ are of sizes (2×3) and (3×2), respectively, then their product $[C]$ can be written as follows:

$$\tag{3.15}$$

In general, if $[A]$ is of size $(l \times m)$ and $[B]$ is of size $(m \times n)$, then the product $[C] = [A][B]$ is of size $(l \times n)$ and its elements c_{ijij} are given by

$$c_{ij} = \sum_k a_{ik} b_{kj} = a_{i1} b_{1j} + a_{i2} b_{2j} + a_{i3} b_{3j} + \cdots \qquad (3.16)$$

The product of two matrices is also a matrix, but matrix multiplication is not commutative.

$$[A][B] \neq [B][A] \qquad (3.17)$$

A matrix may be multiplied by a constant, thusly $c_{ij} = k \cdot a_{ij}$. The result is also a matrix.

4. Inverse Matrix

Though matrix multiplication is possible, the matrix division is not a defined operation. Let $[A]$ be a nonsingular square matrix of order n ; there is another matrix $[A]^{-1}$ called the inverse of $[A]$ for which

$$[A][A]^{-1} = [A]^{-1}[A] = [I] \qquad (3.18)$$

where $[I]$ is the unit matrix of order n. Thus, the multiplication of a matrix with its inverse is analogous to division in the sense that a number divided by itself is equal to 1. The following properties can be applied for the inverse of a square matrix.

(i) if $[A]^{-1}$ exists, it is unique.

(ii) if $[A]^{-1}$ exists, $\left|[A]^{-1}\right| = |A|^{-1} = \dfrac{1}{|A|}$.

(iii) $[A]^{-1}$ exists if and only if $|A| \neq 0$.

(iv) $\left([A]^{-1}\right)^{-1} = [A]$

(v) $[A']^{-1} = \left([A]^{-1}\right)'$, where A' is the transpose of a matrix **A** .

(vi) $\left([A][B]\right)^{-1} = [B]^{-1}[A]^{-1}$

(vii) If $[A]$ is a diagonal matrix, then $[A]^{-1}$ is also a diagonal matrix.

(viii) $[I]^{-1} = [I]$

(ix) If exists, it can be computed as follows:

$$[A]^{-1} = \frac{1}{|A|} \begin{bmatrix} A_{11} & A_{21} & \cdots\cdots & A_{n1} \\ A_{21} & A_{22} & \cdots\cdots & A_{n2} \\ \cdots & \cdots & \cdots & \cdots \\ A_{1n} & A_{2n} & \cdots & A_{nn} \end{bmatrix} \qquad (3.19)$$

where A_{11} A_{21} .... are the cofactors of a_{11}, a_{12} .....in the determinant of the transpose of matrix . The matrix on the right-hand side of Eq. 3.19 is called the adjoint of **A**.

The inverse of a particular matrix may not exist, in which case, the matrix is said to be singular. The solution of a system of simultaneous equations in effect is a problem of evaluating the inverse of a square matrix [4]. Now, we discuss various types of methods for the solution of linear algebraic equations. The methods of solution of the linear algebraic equations may broadly be classified into two categories: direct and iterative methods.

Direct Methods

Direct methods produce the exact solution after a finite number of steps. Iterative methods produce a sequence of approximate solutions, which converges when the numbers of steps tend to infinity [5]. Amongst the direct methods, we will describe matrix inversion, Gauss elimination, LU decomposition, and the Gauss-Jordan method here.

Matrix Inversion Method

The matrix inversion method is a simple direct method used for solving the linear system equations. Recall the system of Eq. 3.2

$$[A][X] = [B] \qquad (3.20)$$

which can be directly solved in the following way. Let $[A]$ be a nonsingular square matrix so that $[A]^{-1}$ exists. Then, multiplying both sides of Eq. 3.20 by $[A]^{-1}$, we obtain:

$$[A][A]^{-1}[X]=[A]^{-1}[B] \tag{3.21}$$

i.e.,
$$[X]=[A]^{-1}[B] \tag{3.22}$$

since
$$[A][A]^{-1}=[I] \text{ and } [I][X]=[X]$$

If $[A]^{-1}$ is known, then the solution vector $[X]$ can be found out from the above matrix relation.

Example 3.1 Solve the system of equations obtained from an electrical network

$$3I_1 + I_2 + 2I_3 = 3$$
$$2I_1 - 3I_2 - I_3 = -3$$
$$I_1 + 2I_2 + I_3 = 4$$

Here,
$$|A| = \begin{vmatrix} 3 & 1 & 2 \\ 2 & -3 & -1 \\ 1 & 2 & 1 \end{vmatrix} = 8$$

also
$$[A]' = \begin{vmatrix} 3 & 2 & 1 \\ 1 & -3 & 2 \\ 2 & -1 & 1 \end{vmatrix}$$

hence,
$$[A]^{-1} = \frac{1}{8}\begin{vmatrix} -1 & 3 & 5 \\ -3 & 1 & 7 \\ 7 & -5 & -11 \end{vmatrix}$$

It follows that

$$[I] = \begin{bmatrix} I_1 \\ I_2 \\ I_3 \end{bmatrix} = \frac{1}{8} \begin{bmatrix} -1 & 3 & 5 \\ -3 & 1 & 7 \\ 7 & -5 & -11 \end{bmatrix} \begin{bmatrix} 3 \\ -3 \\ 4 \end{bmatrix}$$

$$= \frac{1}{8} \begin{bmatrix} 8 \\ 16 \\ -8 \end{bmatrix} = \begin{bmatrix} 1 \\ 2 \\ -1 \end{bmatrix}$$

which gives $I_1 = 1, I_2 = 2$ and $I_3 = -1$.

However, the equations 3.20 are also exactly solvable if the matrix $[A]$ in eq. 3.20 can be transformed into any one of the forms $[D], [L]$ or $[U]$. The matrix inversion method is not suitable for solving large systems as the calculation of $[A]^{-1}$ by cofactors will become difficult. To overcome the difficulties, the following methods can be used for any number of equations.

Gaussian Elimination Method

The Gauss elimination method works systematically, eliminating nonzero elements below the main diagonal of the coefficient matrix. This is accomplished by using only those operations that preserve the solution set of the system, known as elementary row operations:

1. Multiply a row by a nonzero scalar

2. Interchange two rows

3. Multiply a row by a nonzero scalar and add the result to another row.

If we can manipulate one matrix with another using only elementary row operations, then the two matrices are said to be row equivalent.

In order to explain the main concept of the Gauss elimination method, let us consider the following system of three equations in three unknowns:

$$b_{11}x_1 + b_{12}x_2 + b_{13}x_3 = c_1 \tag{3.23a}$$

$$b_{21}x_1 + b_{22}x_2 + b_{23}x_3 = c_2 \tag{3.23b}$$

$$b_{31}x_1 + b_{32}x_2 + b_{33}x_3 = c_3 \qquad \text{(3.23c)}$$

In essence, we wish to eliminate unknowns from the equations by a sequence of algebraic steps. In order to eliminate the unknown variable x_1 from the second equation, we multiply the first equation by $-b_{21}/b_{11}$ and add it to the second equation. In a similar way, we multiply the first equation by $-b_{31}/b_{11}$ and add it to the third equation. Now, we have the following equations:

$$b_{11}x_1 + b_{12}x_2 + b_{13}x_3 = c_1 \qquad \text{(3.24a)}$$

$$b'_{22}x_2 + b'_{23}x_3 = c'_2 \qquad \text{(3.24b)}$$

$$b'_{32}x_2 + b'_{33}x_3 = c'_3 \qquad \text{(3.24c)}$$

Where $b'_{22} = b_{22} - b_{12}\,b_{21}/b_{11}$, $\quad b'_{23} = b_{23} - b_{13}\,b_{21}/b_{11}$, $\quad c'_2 = c_2 - c_1\,b_{21}/b_{11}$, $b'_{32} = b_{32} - b_{12}\,b_{31}/b_{11}$, $b'_{33} = b_{33} - b_{13}\,b_{31}/b_{11}$, $c'_3 = c_3 - c_1\,b_{31}/b_{11}$.

Subsequently, we can eliminate the variable x_2 from the third equation of the system by multiplying the second equation by $-b'_{32}/b'_{22}$ and then adding it to the third equation. Hence, we obtain the following equations:

$$b_{11}x_1 + b_{12}x_2 + b_{13}x_3 = c_1 \qquad \text{(3.25a)}$$

$$b'_{22}x_2 + b'_{23}x_3 = c'_2 \qquad \text{(3.25b)}$$

$$b''_{33}x_3 = c''_3 \qquad \text{(3.25c)}$$

where $b''_{33} = b'_{33} - b'_{23}\,b'_{32}/b'_{22}$, and $c''_3 = c'_3 - c'_2\,b'_{32}/b'_{22}$.

We have eliminated x_1 and x_2 from the third equation and x_1 from the second equation. Now using the backward substitution, solve the third equation for x_3, and substitute in second and first equations. Then, solve the second equation for x_2, and substitute it in the first equation. After that, solve the first equation for x_1.

According to the procedure of elimination of the consecutive unknown variables described above, the elements of these matrices are determined using the following expressions:

$$b_{jk}^{(i+1)} = b_{jk}^{(i)} - \frac{b_{ji}^{(i)}}{b_{ii}^{(i)}} \cdot b_{ik}^{(i)}, \ c_j^{(i+1)} = c_j^{(i)} - \frac{b_{ji}^{(i)}}{b_{ii}^{(i)}} \cdot c_i^{(i)} \tag{3.26}$$

where $i = 1, 2, \ldots\ldots\ldots n$, $j = i+1, i+2, \ldots\ldots\ldots n$ and $k = i+1, i+2, \ldots\ldots\ldots n$;

when $n \geq 3$, the particular terms of the desired solution can be found by using the following recursive formula.

$$x_i^{(m+1)} = \frac{1}{b_{ii}^{(n)}} \left[c_i^{(n)} - \sum_{j=i+1}^{n} b_{ij}^{(n)} \cdot x_j^{(m)} \right] \tag{3.27}$$

where $i = 1, 2, \ldots\ldots n$ and $m = 0, 1, \ldots\ldots\ldots$ with given $x^{(0)}$. The gauss elimination is conducted in $(n-1)$ steps. In the elimination process, if one of the pivot elements b_{11}, b_{22}', or b_{33}'' vanishes or becomes very small compared to other elements in that row, then the method can be modified by rearranging the remaining rows to obtain the nonzero pivot or avoid the multiplication by a large number. This process is called pivoting. If this is not possible, then the matrix becomes singular, and the equations have no solution. However, pivoting is of the following two types: (i) partial pivoting and (ii) complete pivoting.

Partial Pivoting

In the first stage of elimination, the first column is searched for the largest element in magnitude and brought as the first pivot by interchanging the first equation with the equation having the largest element in magnitude. In the second stage of elimination, the second column is searched for the largest element in magnitude among $(n-1)$ elements leaving the first element, and this element is brought as the second pivot by an interchange of the second equation with the equation having the largest element in magnitude. As a result, we can write the following equation to find the pivot:

$$\left| b_{jk}^{(i)} \right| = \max \left| b_{ik}^{(i)} \right|, \ k \leq i \leq n \tag{3.28}$$

and interchange rows k and j.

For example, if during elimination, we have the following situation:

$$x_1 + 2x_2 + 3x_3 = 4$$
$$0.3x_2 + 4x_3 = 5$$
$$-8x_2 + 3x_3 = 6$$

then as $|-8| \succ 0.3$, the second and third equations should be interchanged to yield

$$x_1 + 2x_2 + 3x_3 = 4$$
$$-8x_2 + 3x_3 = 6$$
$$0.3x_2 + 4x_3 = 5$$

It is noted that the interchange of equations does not affect the solution.

Complete Pivoting

We search the matrix $[B]$ for the largest element in magnitude and bring it as the first pivot. This requires not only an interchange of equations but also an interchange of the position of the variables. This leads us to the following algorithm to find the pivot:

Choose l and m as the smallest integers for which

$$\left| b_{lm}^{(i)} \right| = \max \left| b_{ij}^{(i)} \right|, \ k \le i, \ j \le n \tag{3.29}$$

and interchange rows k and l and k and m. If the matrix $[B]$ is diagonally dominant or real, symmetric and positive definite, then no pivoting is necessary.

Gauss-Jordan Elimination Method

Here, the coefficient matrix is reduced to a diagonal matrix rather than a triangular matrix. At all steps of the gauss elimination method, the elimination is done not only in the equations below but also in the equations above, producing the solution without using the back substitution method. On the completion of the Gauss-Jordan method, the equations 3.23a - 3.23c become:A

$$\begin{bmatrix} 1 & 0 & 0 \\ 0 & 1 & 0 \\ 0 & 0 & 1 \end{bmatrix} \begin{bmatrix} x_1 \\ x_2 \\ x_3 \end{bmatrix} = \begin{bmatrix} d_1 \\ d_2 \\ d_3 \end{bmatrix}$$ (3.30)

The solution is given by

$$x_i = d_i, \qquad i = 1(1)n$$

Pseudocode of the Gauss Elimination Method without Partial Pivoting

```
procedure Gauss_Elimination (n,(b_ij),(c_i),(x_i))
real array ((b_ij)_{n×n},(c_i)_n,)
integer i, j, k
for k = 1 to n−1 do
    for i = k+1 to n do
        xmulcoeff ← b_ik / b_kk
        b_ik ← xmulcoeff
        for j = k+1 to n do
            b_ik ← b_ik − (xmulcoeff)·b_kj
        end for
        c_i ← c_i − (xmulcoeff)·c_k
    end for
 end for
x_n ← c_n / b_nn
%% back substitution
for i = n−1 to 1  do
sum  ← c_i
    for j = i+1 to n  do
        sum  ← sum − b_ij·x_j
    end for
    x_i ← sum / b_ii
end for
end procedure Gauss Elimination
```

Generally, it gives a simple method for finding the inverse of a given matrix and the identity matrix I of the same order. However, this method is not used for the solution of a system of equations, and it is more expensive from the computation viewpoint than the Gauss-elimination method.

Example 3.2 Solve the system of three linear equations obtained from an electrical network using (i) Gauss elimination and (ii) Gauss-Jordan method

$$10I_1 - 7I_2 + 0I_3 = 7$$
$$-3I_1 + 2I_2 + 6I_3 = 4$$
$$5I_1 - I_2 + 5I_3 = 6$$

(i) Gauss Elimination Method

According to the Gauss-elimination method, in the first stage, variable I_1 from the second and third equations will be eliminated. To eliminate the unknown variable I_1 from the second equation, we multiply the first equation by $3/10$ and add to the second equation. In a similar way, we multiply the first equation by $-5/10$ and add to the third equation. Now, we have the following equations:

$$10I_1 - 7I_2 + 0I_3 = 7$$
$$-0.1I_2 + 6I_3 = 6.1$$
$$2.5I_2 + 5I_3 = 2.5$$

Here, the pivot in the second equation is very small, so we cannot proceed as usual. We interchange the second and third equations before the second step.

$$10I_1 - 7I_2 = 7$$
$$2.5I_2 + 5I_3 = 2.5$$
$$-0.1I_2 + 6I_3 = 6.1$$

At the second stage, we eliminate I_2 from the third equation by multiplying the second equation by $1/25$ and then adding to the third equation. The resulting system will be upper triangular:

$$10I_1 - 7I_2 = 7$$
$$2.5I_2 + 5I_3 = 2.5$$
$$6.2I_3 = 6.2$$

Back substitution gives the solution:

$$I_1 = 0 \, , I_2 = -1 \text{, and } I_3 = 1$$

(ii) Gauss-Jordan Method

Using the transformation shown in Example 3.2 (i), we obtain the following equations:

$$10I_1 - 7I_2 = 7$$
$$2.5I_2 + 5I_3 = 2.5$$
$$6.2I_3 = 6.2$$

Dividing all three equations by their diagonal elements, we obtain:

$$I_1 - 0.7I_2 = 0.7$$
$$I_2 + 2I_3 = 1$$
$$I_3 = 1$$

Now, we eliminate the variable I_2 from the first equation. To do this, we may add it to the second equation multiplied by 0.7. The resulting equation system has the following form:

$$I_1 - 0 \cdot I_2 + 1.4I_3 = 1.4$$
$$I_2 + 2I_3 = 1$$
$$I_3 = 1$$

Now, we eliminate the variable I_3 from the first and second equations. To do this, multiply the third equation by -1.4 and add it to the first equation. Similarly, the third equation is multiplied by -2 and added to the second equation. Finally, the resulting equations of a system has the following form:

$$I_1 - 0 \cdot I_2 + 0.I_3 = 0$$
$$I_2 + 0.I_3 = -1$$
$$I_3 = 1$$

The transformed matrix coefficients shown here are the diagonal unitary matrices. It follows directly from the equations that $I_1 = 0$, $I_2 = -1$, and $I_3 = 1$. The same results are obtained using the Gauss-elimination method.

LU Matrix Decomposition Method

Let us consider the task of repeatedly solving the system of linear equations:

$$[A][X] = [B] \tag{3.31}$$

each time for the same matrix of coefficients $[A]$, but for different excitation vectors $[B]$. The Gauss elimination and Gauss-Jordan elimination methods are not effective because the repeated transformation of the matrix $[A]$ and vector $[B]$ is needed even though the matrix $[A]$ remains the same always. In such a case, one of the LU decomposition methods, for example, the Crout method [4, 5], may prove to be more convenient. In this method, the decomposition of the nonsingular matrix $[A]$ of the order n into the product $[A] = [L] \cdot [U]$ of the two triangular matrices (lower $[L]$ and upper $[U]$) is used.

We consider, for definiteness, the linear system

$$a_{11}x_1 + a_{12}x_2 + a_{13}x_3 = b_1$$
$$a_{21}x_1 + a_{22}x_2 + a_{23}x_3 = b_2 \tag{3.32}$$
$$a_{31}x_1 + a_{32}x_2 + a_{33}x_3 = b_3$$

which can be written in the form

$$[A][X] = [B] \tag{3.33}$$

let

where

$$[L] = \begin{bmatrix} l_{11} & 0 & 0 \\ l_{21} & l_{22} & 0 \\ l_{31} & l_{32} & l_{33} \end{bmatrix}$$

and

$$[U] = \begin{bmatrix} 1 & u_{12} & u_{13} \\ 0 & 1 & u_{23} \\ 0 & 0 & 1 \end{bmatrix}$$

Hence, Eq. 3.33 becomes

$$[L][U][X] = [B] \tag{3.34}$$

if we set

$$[U] \cdot [X] = [Y] \tag{3.35}$$

then Eq. 3.34 may be written as

$$[L] \cdot [Y] = [B] \tag{3.36}$$

which is equivalent to the system

$$\begin{aligned} l_{11}y_1 &= b_1 \\ l_{21}y_1 + l_{22}y_2 &= b_2 \\ l_{31}y_1 + l_{32}y_2 + l_{33}y_3 &= b_3 \end{aligned} \tag{3.37}$$

and its solution with respect to the vector $[Y] \equiv [y_1, y_2, y_3]^T$ may be obtained without serious difficulties. In the general case $(n > 3)$, the components y_i of the auxiliary vector $[Y]$ can be found by using the following recursive formula:

$$y_1 = b_1 / l_{11} \tag{3.38}$$

$$y_k = \frac{1}{l_{kk}} \left[b_k - \sum_{i=1}^{k-1} l_{ki} \cdot y_i \right], \quad k = 2,3,\ldots\ldots n \tag{3.39}$$

When the column vector $[Y]$ is known, we can solve the matrix equation 3.39, which for n=3 takes the following form.

$$1 \cdot x_1 + u_{12}x_2 + u_{13}x_3 = y_1$$
$$1 \cdot x_2 + u_{23}x_3 = y_2 \qquad\qquad (3.40)$$
$$1 \cdot x_3 = y_3$$

The solution $[X] \equiv [x_1, x_2, x_3]^T$ of these equations can be found in a similarly uncomplicated way, *i.e.*, using the method of consecutive substitutions. It is defined this time by the following recursive computation formula

$$x_j = y_j - \sum_{k=j+1}^{n} u_{jk} \cdot x_i , \quad j = n, n-1, n-2..........1 \qquad (3.41)$$

According to Eqs. 3.39 and 3.41, after substituting the new vector $[B]$, we need to determine only the new vector $[Y]$, and then we must calculate the vector $[X]$, which is the desired solution to our problem. The matrices $[Y][Y]$ do not need to be reprocessed, and this fact diminishes the amount of calculations essentially. It is due to the fact that these matrices were assumed to be known. In the general case, they can be determined using the following recursive relations:

$$l_{i1} = a_{i1} \qquad\qquad (3.42a)$$

$$l_{ij} = a_{ij} - \sum_{k=1}^{j-1} l_{ik} \cdot u_{kj} \qquad\qquad \text{for } i \geq j \geq 1 \qquad (3.42b)$$

$$u_{1j} = \frac{a_{1j}}{l_{11}} \qquad\qquad (3.43a)$$

$$u_{ij} = \frac{1}{l_{ii}} \left[a_{ij} - \sum_{k=1}^{i-1} l_{ik} \cdot u_{kj} \right], \quad \text{for } 1 < i < j \qquad (3.43b)$$

which are in the literature often referred to as Doolittle formulas [9, 10]. The term a_{ij}, where $1 \leq i \leq n$ and $1 \leq j \leq n$ appear in these relations, is the element of a given nonsingular matrix of coefficients $[A]$.

Example 3.3 Solve the equations by using the LU decomposition method.

$$x_1 + 2x_2 + 3x_3 + 4x_4 = 4$$
$$2x_1 + 11x_2 + 20x_3 + 29x_4 = 20.6$$
$$3x_1 + 8x_2 + 16x_3 + 24x_4 = 17.4$$
$$4x_1 + 14x_2 + 25x_3 + 40x_4 = 27.8$$

Using the relations 3.42 and 3.43, we obtain the triangular matrices $[L]$ and $[U]$ as follows:

$$[L] = \begin{bmatrix} 1 & 0 & 0 & 0 \\ 2 & 7 & 0 & 0 \\ 3 & 2 & 3 & 0 \\ 4 & 6 & 1 & 4 \end{bmatrix}$$

and

$$[U] = \begin{bmatrix} 1 & 2 & 3 & 4 \\ 0 & 1 & 2 & 3 \\ 0 & 0 & 1 & 2 \\ 0 & 0 & 0 & 1 \end{bmatrix}$$

The determinant of the matrix of coefficient satisfies the equation det $[A]$=det [L]. det $[U]$=84.1=84. The solution was obtained by using the LU decomposition method $x_1 = 1.0$, $x_2 = 0.7$, $x_3 = 0.4$, $x_4 = 0.1$.

Ill-Conditioned Linear Systems

In practical applications, one usually encounters systems of equations in which small changes in the coefficients of the system produce large changes in the solution. Such systems are said to be ill conditioned. On the other hand, if the corresponding changes in the solution are also small, then the system is well-conditioned. If, from the system

$$[A][X] = [B] \tag{3.44}$$

we obtain the system

$$[C][Y] = [D] \tag{3.45}$$

by introducing small changes in $[A]$ and $[B]$, then the system 3.44 will be ill conditioned when the changes in $[Y]$ are too large compared to those in $[X]$. Otherwise, the system is well conditioned. If the system is ill-conditioned, its coefficient matrix is said to be an ill-conditioned matrix.

Example 3.4. One system

$$2x + y = 2$$
$$2x + 1.01y = 2.01$$

has the solution $x = 1/2$ and $y = 1$, whereas the other system

$$2x + y = 2$$
$$2.01x + y = 2.05$$

has the solution $x = 5$ and $y = -8$, then, the system is ill-conditioned.

Iterative Methods

We have so far discussed some direct methods for the solution of simultaneous linear equations, and we have seen that these methods yield the solution after an amount of computation that is known in advance. We shall now describe the iterative methods, which start from an approximation to the true solution and, if convergent, derive a sequence of closer approximations; the cycle of computations are being repeated till the required accuracy is obtained. This means that in a direct method, the amount of computation is fixed, while in an iterative method, the amount of computations depends on the accuracy required.

In general, one should prefer a direct method for the solution of a linear system, but in the case of matrices with a large number of zero elements, it will be advantageous to use iterative methods to preserve these elements. These methods give a sequence of approximate solutions, which converges when the number of steps tends to infinity. For n > about 40, the one-step methods take too long and accumulate too much round-off error.

Jacobi Method

Let the system be given by

$$a_{11}x_1 + a_{12}x_2 + \ldots\ldots + a_{1n}x_n = b_1$$
$$a_{21}x_1 + a_{22}x_2 + \ldots\ldots + a_{2n}x_n = b_2$$
$$\ldots\ldots\ldots\ldots\ldots\ldots\ldots\ldots\ldots$$
$$a_{m1}x_1 + a_{m2}x_2 + \ldots\ldots + a_{mn}x_n = b_m$$

(3.46)

We assume that the quantities a_{ii} in Eq. 3.46 are pivot elements. The equations may be written as

$$x_1 = \frac{1}{a_{11}}\left(b_1 - \left(a_{12}x_2 + a_{13}x_3 + \ldots\ldots + a_{1n}x_n\right)\right)$$
$$x_2 = \frac{1}{a_{22}}\left(b_2 - \left(a_{21}x_1 + a_{23}x_3 + \ldots\ldots + a_{2n}x_n\right)\right)$$
$$\ldots\ldots\ldots\ldots\ldots\ldots\ldots\ldots\ldots\ldots$$
$$x_n = \frac{1}{a_{mn}}\left(b_m - \left(a_{m1}x_1 + a_{m2}x_2 + \ldots\ldots + a_{mn-1}x_{n-1}\right)\right)$$

(3.47)

Suppose $x_1^{(1)}, x_2^{(1)}, \ldots\ldots x_n^{(1)}$ are any first approximation to the unknown $x_1, x_2, \ldots\ldots x_n$. Substituting in the right side of Eq. 3.47, we find a system of second approximations

$$x_1^{(2)} = \frac{1}{a_{11}}\left(b_1 - \left(a_{12}x_2^{(1)} + a_{13}x_3^{(1)} + \ldots\ldots + a_{1n}x_n^{(1)}\right)\right)$$
$$x_2^{(2)} = \frac{1}{a_{22}}\left(b_2 - \left(a_{21}x_1^{(1)} + a_{23}x_3^{(1)} + \ldots\ldots + a_{2n}x_n^{(1)}\right)\right)$$
$$\ldots\ldots\ldots\ldots\ldots\ldots\ldots\ldots\ldots\ldots$$
$$x_n^{(2)} = \frac{1}{a_{mn}}\left(b_m - \left(a_{m1}x_1^{(1)} + a_{m2}x_2^{(1)} + \ldots\ldots + a_{mn-1}x_{n-1}^{(1)}\right)\right)$$

(3.48)

Similarly, if $x_1^{(n)}, x_2^{(n)}, \ldots\ldots x_n^{(n)}$ are a system of th approximation, then the next approximation is given by the formula

$$x_1^{(n+1)} = \frac{1}{a_{11}}\left(b_1 - \left(a_{12}x_2^{(n)} + a_{13}x_3^{(n)} + \ldots\ldots + a_{1n}x_n^{(n)}\right)\right)$$
$$x_2^{(n+1)} = \frac{1}{a_{22}}\left(b_2 - \left(a_{21}x_1^{(n)} + a_{23}x_3^{(n)} + \ldots\ldots + a_{2n}x_n^{(n)}\right)\right)$$
$$\ldots\ldots\ldots\ldots\ldots\ldots\ldots\ldots\ldots\ldots$$
$$x_n^{(n+1)} = \frac{1}{a_{mn}}\left(b_m - \left(a_{m1}x_1^{(n)} + a_{m2}x_2^{(n)} + \ldots\ldots + a_{mn-1}x_{n-1}^{(n)}\right)\right)$$

(3.49)

In short, we can write the above equation as:

$$x_i^{(m+1)} = \frac{1}{a_{ii}}\left(b_i - \sum_{\substack{j=1 \\ i \neq j}}^{n} a_{ij} x_j^{(m)}\right), \text{ where } i = 1,2,3............n \qquad (3.50)$$

for $m = 0,1,2,..............$, with given $x^{(0)}$. This method is due to Jacobi and is called the method of simultaneous displacements.

Pseudocode of the Jacobi Method

```
procedure Jacobi (A, b, x)
real array ((A)₁ₓₙₓ₁ₓₙ,(b)₁ₙ,(x)₁ₙ)
integer i, j, k max, n
real k max ← 100,  δ ← 10⁻¹⁰, ε ← ½×10⁻⁴
n ← size(A)
for k = 1 to k max do
      y ← x
      for i = 1 to n do
        sum ← bᵢ
        diag ← Aᵢᵢ
        if |diag| < δ   then
           output" diagonal element too small"
       return
      end if
          for j = 1 to n do
        if j ≠ 1   then
            sum ← sum − Aᵢᵢ · yⱼ
         end if
         end for
      xᵢ ← sum/diag
end for
output k, x
if ‖x − y‖ < ε
then output k, x
return
endif
endfor
output" maximum iteration reached!"
return
end Jacobi
```

Example 3.5 Solve the system of linear equations obtained from an electrical network and find out v_1, v_2, and v_3 using the Jacobi method (as shown in Fig. (3.1)).

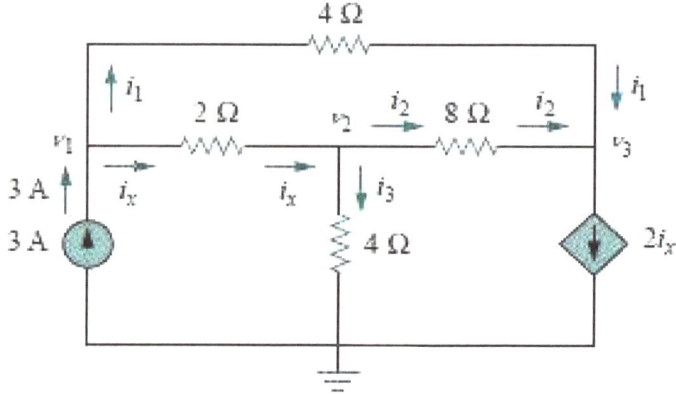

Fig. (3.1). Electrical Network.

At node 1,

Multiplying by 4 and rearranging terms, we get,

At node 2,

Multiplying by 8 and rearranging terms, we get,

At node 3,

Multiplying by 8 and rearranging terms and dividing by 3, we get,

We can solve these three equations using the Jacobi method.

i	v1	v2	v3
1.0000	0	0	0
2.0000	4.0000	0	0
3.0000	4.0000	2.2857	-8.0000
4.0000	2.8571	1.1429	-1.1429
5.0000	4.3810	1.4694	-2.2857
6.0000	4.2177	2.1769	-4.3537
7.0000	4.0000	1.7881	-1.9048
8.0000	4.5572	2.0136	-2.6356
9.0000	4.4639	2.2276	-3.0735
10.0000	4.4605	2.1117	-2.2450
11.0000	4.6595	2.2282	-2.5860
12.0000	4.6235	2.2931	-2.6344
13.0000	4.6506	2.2656	-2.3675
14.0000	4.7212	2.3193	-2.5043
15.0000	4.7114	2.3401	-2.4847

v1 = 4.7982; v2 = 2.3984; v3 = -2.4012

CONVERGENCE ANALYSIS

The standard convergence condition (for any iterative method) is when the spectral radius of the iteration matrix is less than 1:

$$\rho\left(D^{-1}(L+U)\right) < 1 \tag{3.51}$$

A sufficient (but not necessary) condition for the method to converge is that the matrix A is strictly or irreducibly diagonally dominant. Strict row diagonal dominance means that for each row, the absolute value of the diagonal term is greater than the sum of absolute values of other terms:

$$|a_{ii}| > \sum_{j \neq j} |a_{ij}| \tag{3.52}$$

The Jacobi method sometimes converges even if these conditions are not satisfied. Note that the Jacobi method does not converge for every symmetric positive-definite matrix. For example:

$$[A] = \begin{bmatrix} 29 & 2 & 1 \\ 2 & 6 & 1 \\ 1 & 1 & \dfrac{1}{5} \end{bmatrix} \Rightarrow D^{-1}(L+U) = \begin{bmatrix} 0 & \dfrac{2}{29} & \dfrac{1}{29} \\ \dfrac{1}{3} & 0 & \dfrac{1}{6} \\ 5 & 5 & 0 \end{bmatrix} \tag{3.53}$$

$$\Rightarrow \rho\left(D^{-1}(L+U)\right) \approx 1.0661$$

Gauss-Seidel Method

A simple modification of the Jacobi method sometimes yields faster convergence compared to Gauss-Seidel method. In the first equation of 3.47, we substitute the first approximation $x_1^{(1)}, x_2^{(1)}, \ldots \ldots \ldots x_n^{(1)}$ into the right-hand side and denote the result $x_1^{(2)}$. In the second equation, we substitute $x_1^{(2)}, x_2^{(1)}, \ldots \ldots \ldots x_n^{(1)}$ and denote the result $x_2^{(2)}$. In this manner, we complete the first stage of iteration, and the entire process is repeated till the values $x_1, x_2, \ldots \ldots \ldots x_n$ are obtained to the accuracy required. So, we write can write the simultaneous equations for the Gauss-Seidel method as follows:

$$x_1^{(n+1)} = \frac{1}{a_{11}}\left(b_1 - \left(a_{12}x_2^{(n)} + a_{13}x_3^{(n)} + \ldots \ldots + a_{1n}x_n^{(n)}\right)\right)$$

$$x_2^{(n+1)} = \frac{1}{a_{22}}\left(b_2 - \left(a_{21}x_1^{(n+1)} + a_{23}x_3^{(n)} + \ldots \ldots + a_{2n}x_n^{(n)}\right)\right) \tag{3.54}$$

$$\ldots \ldots \ldots \ldots \ldots \ldots \ldots \ldots \ldots \ldots \ldots \ldots \ldots \ldots \ldots$$

$$x_n^{(n+1)} = \frac{1}{a_{mn}}\left(b_m - \left(a_{m1}x_1^{(n+1)} + a_{m2}x_2^{(n+1)} + \ldots \ldots + a_{mn-1}x_{n-1}^{(n+1)}\right)\right)$$

It is clear, therefore, that this method uses an improved component as soon as it is available.

In short, we can write the above equation

$$x_i^{(m+1)} = \frac{1}{a_{ii}}\left(b_i - \sum_{j=1}^{i-1}a_{ij}x_j^{(m+1)} - \sum_{j=i+1}^{n}a_{ij}x_j^{(m)}\right), \text{ where } i = 1,2,3\ldots\ldots\ldots n \quad \textbf{(3.55)}$$

for $m = 0,1,2,\ldots\ldots\ldots$, with given $x^{(0)}$. In the Jacobi method, the equations are solved in order. The components $x_i^{(m)}$ and the corresponding new values $x_i^{(m+1)}$ can be used immediately in their place. The Jacobi and Gauss-Seidel methods converge, for any choice of the first approximation $x_j^{(1)}$ ($j = 1,2,3\ldots\ldots\ldots n$), if every equation of the system 3.48 satisfies the condition that the sum of the absolute values of the coefficients a_{ij}/a_{ii} is almost equal to, or in at least one equation less than unity. The algorithm of the Gauss-Seidel method is given below.

Pseudocode of the Gauss-Seidel Method

```
procedure Gauss_Seid al (A,b,x)
real array ((A)_{1,n,1,n},(b)_{1n},(x)_{1n})
integer i,j,kmax, n
real kmax ← 100,  δ ← 10⁻¹⁰,  ε ← 1/2×10⁻⁴
n ← size(A)
for k = 1 to kmax  do
       y ← x
    for i = 1 to n  do
       sum ← b_i
       diag ← A_{ii}
       if |diag| < δ   then
            output" diagonal element to o small"
            return
       end if
```

for $j = 1$ to $i - 1$ do

    sum $\leftarrow$ sum $- a_{ij} x_j$

end for

for $j = i + 1$ to n do

    sum $\leftarrow$ sum $- a_{ij} \cdot x_j$

end for

$x_i \leftarrow$ sum$/$ *diag*

end for

output k, x

if $\|x - y\| < \varepsilon$ then

    output k, x

    return

endif

endfor

output "max *imum iteration reached*"

return

end Gauss _ Seidal

Example 3.6 Solve the system of linear equations obtained from an electrical network and find out v_1, v_2, and v_3 using Gauss-Seidel method (as shown in Fig. (**3.2**)).

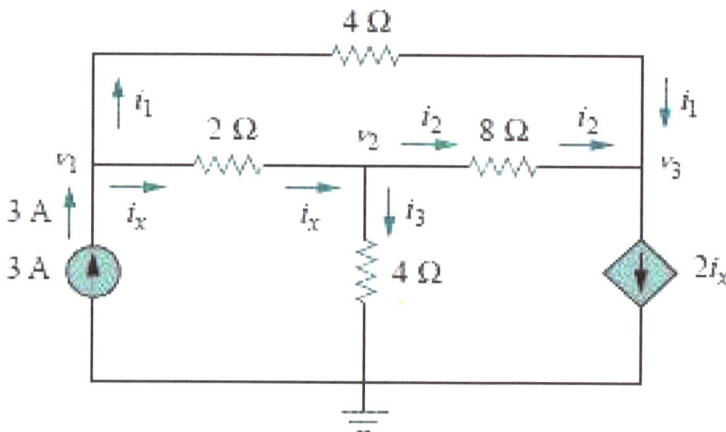

Fig. (3.2). Electrical Network.

At node 1,

$$3 = i_1 + i_x \qquad \Rightarrow \quad 3 = \frac{v_1 - v_3}{4} + \frac{v_1 - v_2}{2}$$

Multiplying by 4 and rearranging terms, we get,

$$3v_1 - 2v_2 - v_3 = 12$$

At node 2,

$$i_x = i_2 + i_3 \qquad \Rightarrow \quad \frac{v_1 - v_2}{2} = \frac{v_2 - v_3}{8} + \frac{v_2 - 0}{4}$$

Multiplying by 8 and rearranging terms, we get,

$$-4v_1 + 7v_2 - v_3 = 0$$

At node 3,

$$i_1 + i_2 = 2i_x \qquad \Rightarrow \quad \frac{v_1 - v_3}{4} + \frac{v_2 - v_3}{8} = \frac{2(v_1 - v_2)}{2}$$

Multiplying by 8 and rearranging terms and dividing by 3, we get,

$$2v_1 - 3v_2 + v_3 = 0$$

We can solve this three equations using Gauss-Seidel method.

i	v1	v2	v3
1.0000	4.7997	0	0
2.0000	4.0000	2.2857	-1.1429
3.0000	5.1429	2.7755	-1.9592
4.0000	5.1973	2.6900	-2.3246
5.0000	5.0185	2.5356	-2.4301
6.0000	4.8804	2.4416	-2.4359
7.0000	4.8158	2.4039	-2.4199
8.0000	4.7960	2.3949	-2.4074
9.0000	4.7941	2.3956	-2.4015

10.0000 4.7966 2.3978 -2.3997

11.0000 4.7987 2.3993 -2.3995

12.0000 4.7997 2.3999 -2.3997

13.0000 4.8000 2.4001 -2.3999

v1 = 4.7997; v2 = 2.3999; v3 = -2.3997.

CONVERGENCE ANALYSIS

The convergence properties of the Gauss-Seidel method are dependent on the matrix A. The procedure is known to converge if either:

- A is symmetric positive-definite, or
- A is strictly or irreducibly diagonally dominant.

The Gauss-Seidel method sometimes converges even if these conditions are not satisfied.

CONSENT FOR PUBLICATION

Not applicable.

CONFLICT OF INTEREST

The author declares no conflict of interest, financial or otherwise.

ACKNOWLEDGEMENT

Declared none.

REFERENCES

[1] M. K, Jain, K. Rajendra, and S. R. K. Iyengar, *Numerical Methods for Scientific and Engineering Computation.* Wiley, 1985.
[2] Jaluria and Yogesh, *Computer Methods for Engineering.* Allyn and Bacon, 1988.
[3] V. Zalizniak, *Essentials of Scientific Computing : Numerical Methods for Science and Engineering.* Chichester, UK: Horwood Pub., 2008.
[4] K. E. Atkinson, *An Introduction to Numerical Analysis.* New York: Wiley, 1978.
[5] J. F. Epperson, *An Introduction to Numerical Methods and Analysis.* Hoboken: Wiley, 2013.

EXERCISES

1. Solve the system of equations obtained from an electrical network using (i) Gauss elimination method, (ii) Gauss-Seidel Method

$$60I_1 - 40I_2 = 200$$
$$-40I_1 + 150I_2 - 100I_3 = 0$$
$$-100I_2 + 130I_3 = 230$$

2. Solve the system of equations obtained from an electrical network using (i) Gauss elimination method, (ii) Gauss-Seidel Method. Verify your answer with Matlab.

$$v_1 - 2v_2 + v_3 = -4$$
$$-2v_1 + 3v_2 + v_3 = 9$$
$$3v_1 + 4v_2 - 5v_3 = 0$$

3. Find the solution of linear equations obtained from an RLC electrical network using (i) Gauss Jordan elimination method, (ii) LU Decomposition Method. Verify your answer with Matlab.

$$5v_1 + 4v_3 + 2v_4 = 3$$
$$v_1 - v_2 + 2v_3 + v_4 = 1$$
$$4v_1 + v_2 + 2v_3 = 1$$
$$v_1 + v_2 + v_3 + v_4 = 0$$

4. Solve the following system of equations using (i) Gauss elimination method, (ii) Jacobi Method (iii) Gauss Seidal Method, and (iv) LU Decomposition method. Verify your answer with Matlab.

$$12v_1 + 7v_2 + 3v_3 = 17$$
$$3v_1 + 6v_2 + 2v_3 = 9$$
$$2v_1 + 7v_2 - 11v_3 = 49$$

Conduct three iterations. Calculate the maximum absolute relative approximate error at the end of each iteration.

5. Establish whether the following matrices are singular, ill conditioned or well-conditioned:

(a)
$$x + y = 2$$
$$x + 1.001y = 2$$

(b)
$$x + y = 0$$
$$0.1x + 1.001y = 0$$

(c)
$$3x_1 - x_2 + 3x_3 = 0$$
$$x_1 + 4x_3 = 0$$
$$2x_1 + 2x_2 + x_3 = 0$$

Fundamentals of Computational Methods for Engineers, 2022, 65-102   **65**

CHAPTER 4

Interpolation, Curve Fitting, and Approximation

Tanvir Ahmed[1*] and **Md. Masud Rana**[1]

[1] *Department of Electrical & Electronic Engineering, Rajshahi University of Engineering & Technology, Bangladesh*

Abstract: In this chapter, we study numerical techniques that deal with given set of data points arising from experimental works. Starting from linear interpolation, different interpolating polynomials are discussed that are used to find functional value at intermediate points of the given data set. Lagrange interpolation is discussed that does not require equally spaced data points. Newton forward and backward difference interpolation formulae are derived to evaluate function near the beginning and end parts of the given data sets. Linear least-squares fit that is widely used to approximate unknown functions is presented and an algorithm is developed. We also discuss least-squares approximations for approximating an explicit function on given interval.

Keywords: Interpolation, Extrapolation, Polynomial curve fitting, Approximation.

INTRODUCTION

In engineering and science, professionally active engineer usually has to work with a large number of numerical data acquired during the calculation, sampling, or experimenting process [1]. These data are the values of a function for a limited number of values of the independent variables. It is often required to interpolate, *i.e.* estimate the value of that function for an intermediate value of the independent variables [2]. For example, data is given only at discrete points such as (x_0, y_0), (x_1, y_1), ...,(x_{n-1}, y_{n-1}), (x_n, y_n). A continuous function $f(x)$ may be used to represent the $n+1$ data values $f(x)$ passing through the $n+1$ points. Then one can find the value of y at any other value of x. This is called interpolation. Of course, if x falls outside the range of x for which the data is given, it is no longer interpolation but instead is called extrapolation. Discrete data points with continuous function f (x) for Interpolation is shown in Fig. (**4.1**).

*****Corresponding author Tanvir Ahmed:** Department of Electrical & Electronic Engineering, Rajshahi University of Engineering & Technology, Bangladesh; E-mail: tanvir_eee_ruet@yahoo.com

Md. Masud Rana, Wei Xu, and Youguang Guo (Eds.)
All rights reserved-© 2022 Bentham Science Publishers

Since we are concerned with interpolation of a discrete function $y_i = y(x_i)$ by means of a continuous function $f(x)$. This problem may be solved in many different ways but the methods most frequently used for this purpose are direct methods, piecewise linear interpolation, using the Lagrange or Newton–Gregory interpolation polynomial, interpolation by means of cubic spline functions and interpolation using a finite linear combination of Chebyshev polynomials of the first kind [3-4]. All these methods are discussed in this chapter.

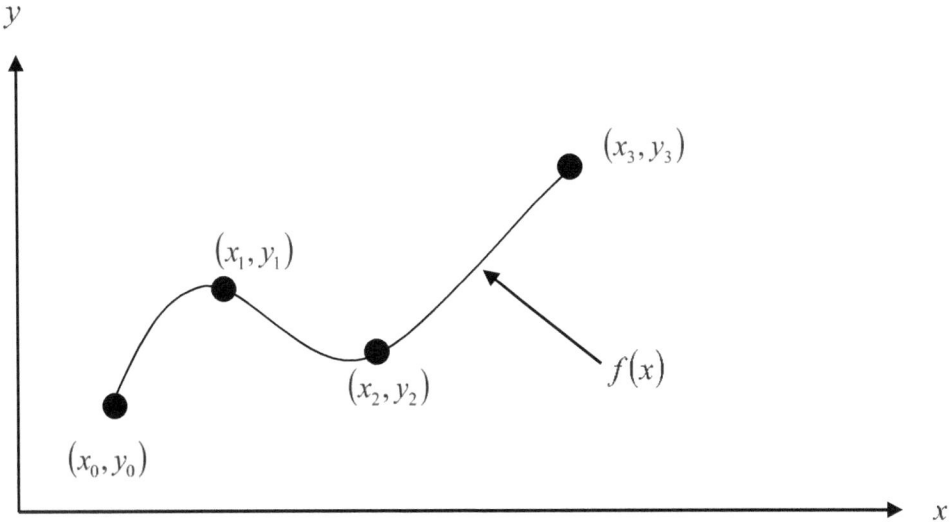

Fig. (4.1). Discrete data points with continuous function $f(x)$ for Interpolation.

DIRECT METHOD

The direct method of interpolation is based on the following premise. Given $n+1$ data points, fit a polynomial of order n as given below

$$y = a_0 + a_1 x + \ldots\ldots\ldots \ldots\ldots + a_n x^n \tag{4.1}$$

through the data, where $a_0, a_1, \ldots\ldots\ldots, a_n$ are $n+1$ real constants. Since $n+1$ values of y are given at $n+1$ values of x, one can write $n+1$ equations. Then the $n+1$ constants $a_0, a_1, \ldots\ldots\ldots, a_n$ can be found by solving the $n+1$ simultaneous linear equations. To find the value of y at a given value of x, simply substitute the value of x in Eq. (4.1).

LINEAR INTERPOLATION

For first-order polynomial interpolation (also called linear interpolation), we choose the value y given by,

$$y(x) = a_0 + a_1 x$$

Given (x_0, y_0), (x_1, y_1), fit a linear interpolation through the data (Fig. **4.2**). Note that $y_0 = f(x_0)$ and $y_1 = f(x_1)$, assuming a linear interpolation means:

$$f_1(x) = a_0 + a_1(x - x_0)$$

Since at $x = x_0$: $f_1(x_0) = f(x_0) = a_0 + a_1(x_0 - x_0) = a_0$,

and at $x = x_1$: $f_1(x_1) = f(x_1) = a_0 + a_1(x_1 - x_0) = f(x_0) + a_1(x_1 - x_0)$

Then

$$a_1 = \frac{f(x_1) - f(x_0)}{x_1 - x_0}$$

so

$$a_0 = f(x_0)$$

$$a_1 = \frac{f(x_1) - f(x_0)}{x_1 - x_0}$$

And the linear interpolant,

$$f_1(x) = a_0 + a_1(x - x_0)$$

Becomes: $f_1(x) = f(x_0) + \dfrac{f(x_1) - f(x_0)}{x_1 - x_0}(x - x_0)$

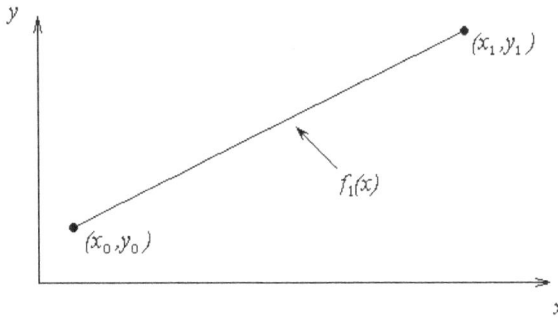

Fig. (4.2). Linear function $f_1(x) = a_0 + a_1(x - x_0)$.

But, it is not necessary to use all the data points. How does one then choose the order of the polynomial and what data points to use? This concept and the direct method of interpolation are best illustrated using examples.

Example 4.1

A robot arm with a rapid laser scanner is doing a quick quality check on holes drilled in a 15"×10" rectangular plate. The centers of the holes in the plate describe the path the arm needs to take, and the hole centers are located on a Cartesian coordinate system (with the origin at the bottom left corner of the plate) given by the specifications in Table **4.1**.

Table 4.1. The coordinates of the holes on the plate.

(in.)	2.00	4.25	5.25	7.81	9.20	10.60
(in.)	7.2	7.1	6.0	5.0	3.5	5.0

If the laser is traversing from $x = 2.00$ to $x = 4.25$ in a linear path, what is the value of y at $x = 4.00$ using the direct method of interpolation and a first-order polynomial?

Solution

For first order polynomial interpolation (also called linear inter polation), we choose the value of y given by

$$y(x) = a_0 + a_1 x$$

Since we want to find the value of y at, using the two points $x_0 = 2.00$

and $x_1 = 4.25$, then

$$x_0 = 2.00, \; y(x_0) = 7.2$$

$$x_1 = 4.25, \; y(x_1) = 7.1$$

gives

$$y(2.00) = a_0 + a_1(2.00) = 7.2$$

$$y(4.25) = a_0 + a_1(4.25) = 7.1$$

Writing the equations in matrix form, we have

$$\begin{bmatrix} 1 & 2.00 \\ 1 & 4.25 \end{bmatrix} \begin{bmatrix} a_0 \\ a_1 \end{bmatrix} = \begin{bmatrix} 7.2 \\ 7.1 \end{bmatrix}$$

Solving the above two equations gives

$$a_0 = 7.2889$$

$$a_1 = -0.044444$$

Hence

$$y(x) = a_0 + a_1 x$$

$$y(x) = 7.2889 - 0.044444 \, x, \quad 2.00 \le x \le 4.25$$

$$y(4.00) = 7.2889 - 0.044444 \, (4.00)$$

$$= 7.1111 \text{ in.}$$

QUADRATIC INTERPOLATION

Given (x_0, y_0), (x_1, y_1), and (x_2, y_2), fit a quadratic interpolant through the data.

Note that $y = f(x)$, $y_0 = f(x_0)$, $y_1 = f(x_1)$, and $y_2 = f(x_2)$, assume the quadratic

interpolant $f_2(x)$ given by

$$f_2(x) = a_0 + a_1(x - x_0) + a_2(x - x_0)(x - x_1)$$

At $x = x_0$

$$f(x_0) = f_2(x_0) = a_0 + a_1(x_0 - x_0) + a_2(x_0 - x_0)(x_0 - x_1)$$

$$= a_0$$

$$a_0 = f(x_0)$$

At $x = x_1$

$$f(x_1) = f_2(x_1) = a_0 + a_1(x_1 - x_0) + a_2(x_1 - x_0)(x_1 - x_1)$$

$$f(x_1) = f(x_0) + a_1(x_1 - x_0)$$

then

$$a_1 = \frac{f(x_1) - f(x_0)}{x_1 - x_0}$$

At $x = x_2$

$$f(x_2) = f_2(x_2) = a_0 + a_1(x_2 - x_0) + a_2(x_2 - x_0)(x_2 - x_1)$$

$$f(x_2) = f(x_0) + \frac{f(x_1) - f(x_0)}{x_1 - x_0}(x_2 - x_0) + a_2(x_2 - x_0)(x_2 - x_1)$$

then

$$a_2 = \frac{\dfrac{f(x_2) - f(x_1)}{x_2 - x_1} - \dfrac{f(x_1) - f(x_0)}{x_1 - x_0}}{x_2 - x_0}$$

Hence the quadratic interpolant is given by

$$f_2(x) = a_0 + a_1(x - x_0) + a_2(x - x_0)(x - x_1)$$

$$f_2(x) = f(x_0) + \frac{f(x_1) - f(x_0)}{x_1 - x_0}(x - x_0) + \frac{\dfrac{f(x_2) - f(x_1)}{x_2 - x_1} - \dfrac{f(x_1) - f(x_0)}{x_1 - x_0}}{x_2 - x_0}(x - x_0)(x - x_1)$$

Example 4.2

A robot arm with a rapid laser scanner is doing a quick quality check on holes drilled in a 15"×10" rectangular plate. The centers of the holes in the plate describe the path the arm needs to take, and the hole centers are located on a Cartesian coordinate system (with the origin at the bottom left corner of the plate) given by the specifications in Table **2**.

Table 4.2. The coordinates of the holes on the plate.

(in.)	2.00	4.25	5.25	7.81	9.20	10.60
(in.)	7.2	7.1	6.0	5.0	3.5	5.0

If the laser is traversing from $x = 2.00$ to $x = 4.25$ $x = 5.25$ in a quadratic path, what is the value of y at $x = 4.00$ using the direct method of interpolation and a second-order polynomial? Find the absolute relative approximate error for the second-order polynomial approximation.

Solution

For second-order polynomial interpolation (also called quadratic interpolation), we choose the value of y from the Fig. (**4.3**),

$$y(x) = a_0 + a_1 x + a_2 x^2$$

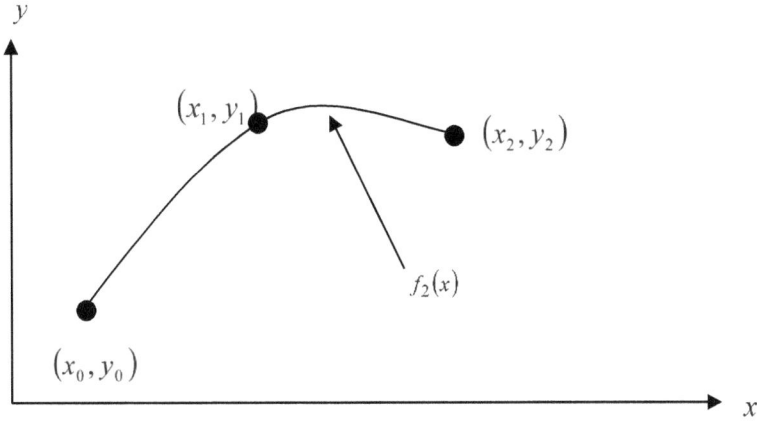

Fig. (4.3). Quadratic interpolation.

Since we want to find the value of y at, using the three points as

$x_0 = 2.00$, $x_1 = 4.25$ and $x_2 = 5.25$, then

$$x_0 = 2.00, \ y(x_0) = 7.2$$

$$x_1 = 4.25, \ y(x_1) = 7.1$$

$$x_2 = 5.25, \ y(x_2) = 6.0$$

gives

$$y(2.00) = a_0 + a_1(2.00) + a_2(2.00)^2 = 7.2$$

$$y(4.25) = a_0 + a_1(4.25) + a_2(4.25)^2 = 7.1$$

$$y(5.25) = a_0 + a_1(5.25) + a_2(5.25)^2 = 6.0$$

Writing the three equations in matrix form, we have

$$\begin{bmatrix} 1 & 2.00 & 4 \\ 1 & 4.25 & 18.063 \\ 1 & 5.25 & 27.563 \end{bmatrix} \begin{bmatrix} a_0 \\ a_1 \\ a_2 \end{bmatrix} = \begin{bmatrix} 7.2 \\ 7.1 \\ 6.0 \end{bmatrix}$$

Solving the above three equations gives

$a_0 = 4.5282$

$a_1 = 1.9855$

$a_2 = -0.32479$

Hence

$y(x) = 4.5282 + 1.9855 x - 0.32479 x^2, \; 2.00 \le x \le 5.25$

At   ,

$y(4.00) = 4.5282 + 1.9855(4.00) - 0.32479(4.00)^2$

$= 7.2735$ in.

The absolute relative approximate error $|\in_a|$ obtained between the results from the first and second-order polynomial is

$$|\in_a| = \left| \frac{7.2735 - 7.1111}{7.2735} \right| \times 100$$

$= 2.2327\%$

Similarly, for third-order polynomial interpolation (also called cubic interpolation), we choose the value of y given by:

$$y(x) = a_0 + a_1 x + a_2 x^2 + a_3 x^3$$

The same process as described before can be applied to obtain the coefficients. Now the other interpolation method is described as follows.

LAGRANGE INTERPOLATION

Consider the following table as an example of known function for Lagrange interpolation. Let compute the value of ln 2.5.

x	2	3	4	5
$y = \ln x$	0.6931	1.0986	1.3863	1.6094

Consider the simplest polynomial of the following form to find the value of y for x = 2.5

$$f(x) = c_1 + c_2 x \qquad (4.2)$$

At $x = x_1 = 2$ and $x = x_2 = 3$ we have

$$f(x_1) = \ln x_1 = c_1 + c_2 x_1 = 0.6931$$
$$f(x_2) = \ln x_2 = c_1 + c_2 x_2 = 1.0986$$

These two equations can be solved for c_1 and c_2, which result as follows

$$c_2 = \frac{f(x_1) - f(x_2)}{x_1 - x_2} = 0.4055$$
$$c_1 = f(x_1) - c_2 x_1 = -0.1178$$

Therefore, from Eq. (4.2) one can obtain the value of y for $x = 2.5$

$$f(2.5) = -0.1178 + 0.4055 \times 2.5 = 0.8959$$

The actual value of $\ln 2.5$ is known to be 0.9163. Therefore linear approximation of the polynomial is not suitable. Let assume the second order polynomial as follows

$$f(x) = c_1 + c_2 x + c_3 x^2$$

This form of the second order polynomial is not suitable to calculate the coefficients c_1, c_2 and c_3. Therefore, we consider the following polynomial to simplify the calculation of the coefficient.

$$f(x) = c_1 + c_2(x - x_1) + c_3(x - x_1)(x - x_2)$$

The coefficients c_1, c_2 and c_3, can be obtained from the following simultaneous equations

$$f(x_1) = c_1$$
$$f(x_2) = c_1 + c_2(x_2 - x_1)$$
$$f(x_3) = c_1 + c_2(x_3 - x_1) + c_3(x_3 - x_1)(x_3 - x_2)$$

Thus we have

$$c_1 = f(x_1) = 0.6931$$

$$c_2 = \frac{f(x_2) - c_1}{(x_2 - x_1)} = \frac{f(x_2) - f(x_1)}{(x_2 - x_1)} = 0.4055$$

$$c_3 = \frac{f(x_3) - f(x_1) - \dfrac{f(x_2) - f(x_1)}{(x_2 - x_1)}(x_3 - x_1)}{(x_3 - x_1)(x_3 - x_2)} = -0.0589$$

Therefore,

$$f(2.5) = 0.6931 + 0.4055 \times (2.5 - 2) - 0.0589 \times (2.5 - 2)(2.5 - 3) = 0.9106$$

This value of $f(x)$ is close to the actual value of ln 2.5. The coefficient calculation can be simplified more by considering the following form of the polynomial

$$f(x) = c_1(x - x_2)(x - x_3) + c_2(x - x_1)(x - x_3) + c_3(x - x_1)(x - x_2)$$

Thus the coefficients of the polynomial $f(x)$ can be obtained without solving simultaneous equations as follows:

$$c_1 = \frac{f(x_1)}{(x_1 - x_2)(x_1 - x_3)}$$

$$c_2 = \frac{f(x_2)}{(x_2 - x_1)(x_2 - x_3)}$$

$$c_3 = \frac{f(x_3)}{(x_3 - x_1)(x_3 - x_2)}$$

Therefore, the polynomial $f(x)$ can be written as

$$f(x) = \frac{f(x_1)(x - x_2)(x - x_3)}{(x_1 - x_2)(x_1 - x_3)} + \frac{f(x_2)(x - x_1)(x - x_3)}{(x_2 - x_1)(x_2 - x_3)} + \frac{f(x_3)(x - x_1)(x - x_2)}{(x_3 - x_1)(x_3 - x_2)}$$

Or in the general form

$$f(x) = \sum_{i=1}^{n+1} f(x_i) \prod_{\substack{j \neq i \\ j=1}}^{n+1} \frac{(x - x_i)}{(x_j - x_i)} \tag{4.3}$$

where n is the order of the polynomial $f(x)$. The polynomial in Eq. (4.3) is known as Lagrange interpolation polynomial. It is easy to program to interpolate based on the Lagrange interpolation polynomial. However, hand calculation is cumbersome. Let consider the following table and observe the effect of increasing the order of the Lagrange interpolation polynomial.

x	0.1	0.2	0.3	0.4	0.5
$y = \cosh x$	1.0050	1.0201	1.0453	1.0811	1.1276

Suppose we need to find the value of $y = \cosh x$ *at* $x = 3.5$. Second-order Lagrange polynomial leads to

$$f(x) = \frac{f(x_1)(x-x_2)(x-x_3)}{(x_1-x_2)(x_1-x_3)} + \frac{f(x_2)(x-x_1)(x-x_3)}{(x_2-x_1)(x_2-x_3)} + \frac{f(x_3)(x-x_1)(x-x_2)}{(x_3-x_1)(x_3-x_2)}$$

$$f(3.5) = \frac{1.0050 \times (3.5-0.2)(3.5-0.3)}{(0.1-0.2)(0.1-0.3)} + \frac{1.0201 \times (3.5-0.1)(3.5-0.3)}{(0.2-0.1)(0.2-0.3)}$$

$$+ \frac{1.0453 \times (3.5-0.1)(3.5-0.2)}{(0.3-0.1)(0.3-0.2)}$$

$$= 0.3769 - 1.2751 + 1.9599$$

$$= 1.0617$$

The true value of cosh 3.5 is found as 1.0619. Third-order Lagrange polynomial leads to $f(x) = 1.0618$ and fourth-order Lagrange polynomial results in $f(x) = 1.0619$, which is equal to the true value of cosh 3.5. However, higher-order polynomial may not confirm better accuracy always. One major advantage of the Lagrange Interpolation is that the data points of x do not need to be equidistant. An algorithm for Lagrange Interpolation to interpolate is given below:

Algorithm for Lagrange Interpolation

1. Read x, n

2. *for* $i = 1$ to $n+1$ in steps of 1 *do*

3. Read x_i, fx_i

4. *endfor*

5. sum $\leftarrow 0$

6. *for* $i = 1$ to $n+1$ in steps of 1 *do*

7. product $\leftarrow 1$

8. *for* $j = 1$ to $n+1$ in steps of 1 *do*

9. *if* $(j \neq i)$ then

10. product $\leftarrow$ product $\times (x - x_j)/(x_i - x_j)$

11. *endfor*

12. sum $\leftarrow$ sum $+$ product $\times fx_i$

13. *endfor*

14. Write sum

15. Stop

NEWTON INTERPOLATION

Consider the following polynomial

$$f(x) = y_n(x) = c_0 + c_1(x - x_0) + c_2(x - x_0)(x - x_1)$$
$$+ c_3(x - x_0)(x - x_1)(x - x_2) + \ldots \qquad \textbf{(4.4a)}$$
$$+ c_n(x - x_0)(x - x_1)(x - x_2)\ldots(x - x_{n-1})$$

Substituting the tabulated values of x in the above equation leads to the determination of the coefficients $c_0, c_1, c_2, \ldots, c_n$. Therefore, we have

$$c_0 = y_0(x_0) = y_0;$$
$$y_1(x_1) = y_1 = c_0 + c_1(x_1 - x_0)$$
$$\Rightarrow c_1 = \frac{y_1 - y_0}{x_1 - x_0} = \frac{\Delta y_0}{h};$$

$$y_2 = c_0 + c_1(x_2 - x_0) + c_2(x_2 - x_0)(x_2 - x_1) = y_0 + \frac{y_1 - y_0}{h} 2h + c_2 \times 2h \times h$$
$$\Rightarrow c_2 = \frac{y_2 - 2y_1 + y_0}{h^2 2!} = \frac{(y_2 - y_1) - (y_1 - y_0)}{h^2 2!} = \frac{\Delta y_1 - \Delta y_0}{h^2 2!} = \frac{\Delta^2 y_0}{h^2 2!};$$

Similarly, $c_3 = \dfrac{\Delta^3 y_0}{h^3 3!}, c_4 = \dfrac{\Delta^4 y_0}{h^4 4!}, \ldots, c_n = \dfrac{\Delta^n y_0}{h^n n!}$.

Substituting the value of coefficients into the Eq. (4.4), we have

$$f(x) = y_0 + \frac{\Delta y_0}{h}(x - x_0) + \frac{\Delta^2 y_0}{h^2 2!}(x - x_0)(x - x_1)$$

$$+ \frac{\Delta^3 y_0}{h^3 3!}(x - x_0)(x - x_1)(x - x_2) + \ldots \qquad \textbf{(4.4 b)}$$

$$+ \frac{\Delta^n y_0}{h^n n!}(x - x_0)(x - x_1)(x - x_2)\ldots(x - x_{n-1})$$

Suppose $x = x_0 + kh$, so $k = (x - x_0)/h$. Therefore, Eq. (4.4) leads to

$$f(x) = y_0 + k\Delta y_0 + k(k-1)\frac{\Delta^2 y_0}{2!} + k(k-1)(k-2)\frac{\Delta^3 y_0}{3!}$$

$$+ \ldots + k(k-1)(k-2)\cdots(k-n+1)\frac{\Delta^n y_0}{n!}. \qquad \textbf{(4.5)}$$

The Eq. (4.5) is known as Newton's forward difference interpolation formula. This formula is useful to interpolate towards the beginning of a set of tabular values. However, to interpolate near the end of the tabular values, one needs to use Newton's backward difference interpolation formula. We can arrive at Newton's backward difference interpolation formula assuming an nth order polynomial in the following form

$$f(x) = y_n(x) = c_0 + c_1(x - x_n) + c_2(x - x_n)(x - x_{n-1})$$

$$+ c_3(x - x_n)(x - x_{n-1})(x - x_{n-2}) + \ldots \qquad \textbf{(4.6)}$$

$$+ c_n(x - x_n)(x - x_{n-1})(x - x_{n-2})\ldots(x - x_1).$$

Substituting the tabulated values $(x_n, x_{n-1}, x_{n-2}, \ldots, x_2, x_1, x_0)$ of x in Eq. (4.6) leads to the determination of the coefficients $c_0, c_1, c_2, \ldots, c_n$.

$$c_0 = y_n;$$

$$c_1 = \frac{y_n - y_{n-1}}{x_n - x_{n-1}} = \frac{\nabla y_n}{h};$$

$$c_2 = \frac{y_n - 2y_{n-1} + y_{n-2}}{h^2 2!} = \frac{\nabla y_n - \nabla y_{n-1}}{h^2 2!} = \frac{\nabla^2 y_n}{h^2 2!};$$

$$c_3 = \frac{\nabla^3 y_n}{h^3 3!}, c_4 = \frac{\nabla^4 y_n}{h^4 4!}, \cdots, c_n = \frac{\nabla^n y_n}{h^n n!}.$$

Substituting the value of coefficients into the Eq. (4.6), we have

$$f(x) = y_n + \frac{\nabla y_n}{h}(x - x_n) + \frac{\nabla^2 y_n}{h^2 2!}(x - x_n)(x - x_{n-1})$$

$$+ \frac{\nabla^3 y_n}{h^3 3!}(x - x_n)(x - x_{n-1})(x - x_{n-2}) + \cdots$$

$$+ \frac{\nabla^n y_n}{h^n n!}(x - x_n)(x - x_{n-1})(x - x_{n-2})\cdots(x - x_1)$$

Therefore, the Newton's backward difference interpolation formula results as

$$f(x) = y_n + k\nabla y_n + k(k+1)\frac{\nabla^2 y_n}{2!} + k(k+1)(k+2)\frac{\nabla^3 y_n}{3!}$$

$$+ \cdots + k(k+1)(k+2)\cdots(k+n-1)\frac{\nabla^n y_n}{n!},$$

(4.7)

where $k = (x - x_n)/h$.

Example 4.3.

The tabular data points in the following table are obtained from a second-degree polynomial $y(x) = 10 + 2x + 4x^2$. Find the value of $y(11)$ using Newton's forward and backward difference interpolation and comment on the result. Note that point $y(11)$ is outside the tabulated points. The method by which the value of y is found other than the given range is known as extrapolation.

X	1	2	3	4	5	6	7	8	9	10
Y	16	30	52	82	120	166	220	282	352	430

Solution

Let form the difference table.

x	y		2	3
1	16			
		14		
2	30		8	
		22		0
3	52		8	
		30		0
4	82		8	
		38		0
5	120		8	
		46		0
6	166		8	
		54		0
7	220		8	
		62		0
8	282		8	
		70		0
9	352		8	
		78		
10	430			

From Eq. (4.5) the forward interpolation formula is

$$y(x) = y_0 + k\Delta y_0 + k(k-1)\frac{\Delta^2 y_0}{2!} + k(k-1)(k-2)\frac{\Delta^3 y_0}{3!}$$

Substituting $x_0 = 1$ and $h = 1$ in $x = x_0 + kh$, $k = x - 1$. Thus,

$$y(x) = 16 + (x-1)14 + (x-1)(x-2)\frac{8}{2} + 0$$

$$y(x) = 10 + 2x + 4x^2$$

Substituting $x_n = 10$, $k = (x\text{-}x_n)/h = x - 10$. From Eq. (4.6) the backward interpolation formula is

$$y(x) = y_n + k\nabla y_n + k(k+1)\frac{\nabla^2 y_n}{2!} + k(k+1)(k+2)\frac{\nabla^3 y_n}{3!}$$

$$y(x) = 430 + (x-10)78 + (x-10)(x-9)\frac{8}{2} + 0$$

$$y(x) = 10 + 2x + 4x^2$$

The difference table shows that the third differences are zero and the second differences are constant. So, the tabulated points represent a second degree polynomial. Since both the forward and backward interpolation yield same polynomial. Therefore, both the forward and backward interpolation formula can be used for interpolation and extrapolation.
So,

$$y(11) = 10 + 2 \times 11 + 4 \times 11^2 = 516$$

Example 4.4.

The tabulated points below shows the value of $y(x) = \sin x$ in the range $0.1 \leq x \leq 0.5$. Compare the forward and backward interpolation formula for interpolation and extrapolation.

x	0.1	0.2	0.3	0.4	0.5
y	0.0998	0.1987	0.2955	0.3894	0.4794

Solution

Let form the difference table.

x	y	Δ	Δ^2	Δ^3	Δ^4
0.1	0.0998	-	-	-	-
-	-	0.098835914	-	-	-
0.2	0.1987	-	-0.00199	-	-
-	-	0.096850876	-	-0.00097	-
0.3	0.2955		-0.00295	-	0.00003
-	-	0.093898136	-	-0.00094	-
0.4	0.3894	-	-0.00389	-	-
-	-	0.090007196	-	-	-
0.5	0.4794	-	-	-	-

Let compare the forward and backward interpolation formula for interpolation and extrapolation

x	Actual Value of sin x	Forward interpolation	Backward interpolation
0.01	0.01000	0.01088	0.03853
0.1	0.09983	0.09983	0.11949
0.15	0.14944	0.14925	0.1644
0.2	0.19867	0.19867	0.20945
0.25	0.24740	0.24809	0.25444
0.3	0.29552	0.29751	0.29943
0.35	0.34290	0.34693	0.34442
0.4	0.38942	0.39635	0.38942
0.45	0.43497	0.44578	0.43442
0.5	0.47943	0.49520	0.47943
0.55	0.52269	0.54463	0.52443
0.6	0.56464	0.59405	0.56945
0.7	0.64422	0.69291	0.65949

The table demonstrates that forward interpolation is useful to get the values of $y(x)$ near the beginning of the range $0.1 \leq x \leq 0.5$ whereas the backward interpolation yields satisfactory results at the end of the range $0.1 \leq x \leq 0.5$. Extrapolation (if the tabulated data points do not represent a polynomial) very far away from the range $0.1 \leq x \leq 0.5$ may be dangerous.

CURVE FITTING TECHNIQUE

In many engineering applications, experimental and numerical results are presented in the form of a tabulated data set of finite numbers. It is often required to find the value of dependent variables at intermediate points where tabulated data points are not available [5]. A useful technique is to represent the data   set by means of a continuous mathematical function that best fit the given set of data points. This procedure is known as curve fitting.

Curve fitting is applied to solve a broad range of engineering problems. Experimental data, such as the variation of electrical current in a load, motor speed as a function of applied voltage, wind speed at various heights above the ground surface at different places, pressure or temperature variation with time, are used to find a continuous mathematical function to best fit the data. This function can be used to solve relevant engineering problems.

To obtain best fit to a given data set, the function is chosen based on the physical nature of the given data. For example, sinusoidal functions can be employed to fit data that represent natural phenomena vary periodically. The voltage across the resistor of a dc RC circuit varies exponentially with time. Therefore, exponential functions can be employed for curve fitting to find any voltage from the data set. Appropriate choice of the function for curve fitting provides the preferred result in a simple and useful form and reduces the number of unknown parameters. There are several numerical methods for curve fitting to a given data set. The most extensively used method is least-squares fit.

LINEAR LEAST SQUARES FIT

Linear least-squares fit is a method to approximate an unknown function $y = f(x)$ expressed in the tabular form obtained from experimental results by a linear function $\phi(x) = C_0 + C_1 x$. The procedure to find the unknown coefficients C_0 and C_1 to best fit the given data set is known as linear least-squares regression. As the experimental data contains error, it is not necessary to fit a curve that passes through every data points. Instead, a curve is required that is a close approximation of the experimental data and represents the general trend of the data. The least square method fit a unique curve systematically through the given data points. In this method, a curve is obtained such that the sum of the squares of the distances between the data points and the curve becomes minimum.

Let the set of given data points be (x_i, y_i), $i = 1, 2, \ldots, n$, where x is the independent variable and $y = f(x)$ is the dependent variable. The function to fit the data points is $\phi(x) = C_0 + C_1 x$, where C_0 and C_1 are unknown coefficients that need to be determined. The sum of the squares of the distances of the data from the polynomial $\phi(x)$ is given by:

$$S(C_0, C_1) = \sum_{i=1}^{n} [f(x_i) - (C_0 + C_1 x_i)]^2$$

In the least square method, the sum of the square of distances must be minimum which depends on C_0 and C_1. Therefore, differentiating the above equations with respect to C_0 and C_1 and equating to zero results in

$$\frac{\partial S}{\partial C_0} = 2 \sum_{i=1}^{n} [f(x_i) - (C_0 + C_1 x_i)](-1) = 0$$

$$\frac{\partial S}{\partial C_1} = 2 \sum_{i=1}^{n} [f(x_i) - (C_0 + C_1 x_i)](-x_i) = 0$$

Rearranging the above equations leads to the following linear simultaneous equations for C_0 and C_1.

$$C_0 n + C_1 \sum_{i=1}^{n} x_i = \sum_{i=1}^{n} f(x_i)$$

$$C_0 \sum_{i=1}^{n} x_i + C_1 \sum_{i=1}^{n} x_i^2 = \sum_{i=1}^{n} x_i f(x_i)$$

The above equations are called normal equations. The solutions of coefficients C_0 and C_1 are

$$C_0 = \frac{\sum_{i=1}^{n} f(x_i) \sum_{i=1}^{n} x_i^2 - \sum_{i=1}^{n} x_i f(x_i) \sum_{i=1}^{n} x_i}{n \sum_{i=1}^{n} x_i^2 - (\sum_{i=1}^{n} x_i)^2}$$

$$C_1 = \frac{n \sum_{i=1}^{n} x_i f(x_i) - \sum_{i=1}^{n} f(x_i) \sum_{i=1}^{n} x_i}{n \sum_{i=1}^{n} x_i^2 - (\sum_{i=1}^{n} x_i)^2}$$

An algorithm to fit data points based on the linear least-squares method is given below:

Algorithm for Linear least-squares fit

1. Read n
2. sum $x \leftarrow 0$
3. sum $x\_square \leftarrow 0$
4. sum $y \leftarrow 0$
5. sum $xy \leftarrow 0$
6. *for* $i = 1$ to n *do*
7. read x, y
8. sum $x \leftarrow$ sum $x + x$
9. sum $y \leftarrow$ sum $y + y$
10. sum $x\_square \leftarrow$ sum $x\_square + x^2$
11. sum $xy \leftarrow$ sum $xy + x \times y$
12. *endfor*
13. demoninator $\leftarrow n \times$ sum x_square - sum $x \times$ sum x
14. $C_0 \leftarrow ($ sum $y \times$ sum x_square - sum $xy \times$ sum $x)/$ demoninator
15. $C_1 \leftarrow (n \times$ sum xy- sum $y \times$ sum $x)/$ demoninator
16. Write C_0 and C_1
17. Stop

Example 4.5.

The table below gives the temperature T (in °C) change in a chamber with time t (in minutes). Assume $T = C_0 + C_1 t$, find the best values of C_0 and C_1 using least-squares method.

t (in m)	10	20	30	40	50	60	70	80	90	100
T(in °C)	19	22	25	28	30	31	35	37	39	42

Solution

To fit the data points using lest square method, consider the following table:

t	T	t^2	t^3	tT
10	19	100	1000	190
20	22	400	8000	440
30	25	900	27000	750
40	28	1600	64000	1120
50	30	2500	125000	1500
60	31	3600	216000	1860
70	35	4900	343000	2450
80	37	6400	512000	2960
90	39	8100	729000	3510
100	42	10000	1000000	4200
$\sum t = 550$	$\sum T = 308$	$\sum t^2 = 38500$	$\sum t^3 = 3025000$	$\sum tT = 18980$

For linear least squares method linear simultaneous equations are

$$C_0 n + C_1 \sum_{i=1}^{n} t_i = \sum_{i=1}^{n} T_i$$

$$C_0 \sum_{i=1}^{n} t_i + C_1 \sum_{i=1}^{n} t_i^2 = \sum_{i=1}^{n} t_i T_i$$

Therefore,

$$10 C_0 + 550 C_1 = 308$$

$$550C_0 + 3025000C_1 = 18980$$

From which we obtain, $C_0 = 17.2000$ and $C_1 = 0.2473$. Fig. (**4.4**) shows linear least squares curve fitting graphically .

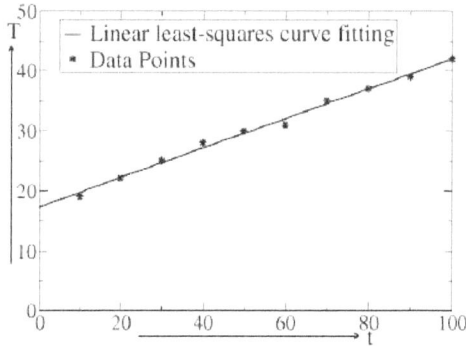

Fig. (4.4). Example showing linear least squares curve fitting.

POLYNOMIAL CURVE FITTING

The linear curve fitting technique is easy to apply because only two coefficients C_0 and C_1 need to be determined. In many cases, the trend of the experimental data is hardly represented by a straight line. In such case, nonlinear function curve fitting is more satisfactory compared to the curve fitting with a linear function. For example, the data points in Fig. (**4.5**) have the trend of a parabola which is a second-degree polynomial. This section illustrates curve fitting by a kth degree polynomial.

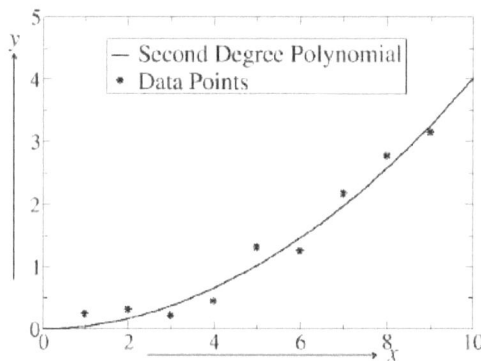

Fig. (4.5). Trend of data pints and polynomial curve fitting.

Let the set of given data points be (x_i, y_i), $i = 1,2,\ldots, n$, and the function to fit the data points is given by k^{th} degree polynomial $\phi(x) = C_0 + C_1 x + C_2 x^2 + \cdots + C_k x^k$. The sum of the squares of the distances of the data point from the polynomial $\phi(x)$ is given by:

$$S(C_0, C_1, C_2 \ldots C_k) = \sum_{i=1}^{n} [f(x_i) - \phi(x)]^2$$

$$S(C_0, C_1, C_2 \ldots C_k) = \sum_{i=1}^{n} [f(x_i) - (C_0 + C_1 x_i + C_2 x_i^2 + \cdots + C_k x_i^k)]^2$$

In the least square approximation method, the sum of the square of distances must be minimum which depends on $C_0, C_1, C_2 \ldots C_k$. Let's find the value of unknown coefficients $C_0, C_1, C_2 \ldots C_k$.

$$\frac{\partial S}{\partial C_0} = 2 \sum_{i=1}^{n} [f(x_i) - (C_0 + C_1 x_i + C_2 x_i^2 + \cdots + C_k x_i^k)](-1) = 0$$

$$\frac{\partial S}{\partial C_1} = 2 \sum_{i=1}^{n} [f(x_i) - (C_0 + C_1 x_i + C_2 x_i^2 + \cdots + C_k x_i^k)](-x_i) = 0$$

$$\frac{\partial S}{\partial C_2} = 2 \sum_{i=1}^{n} [f(x_i) - (C_0 + C_1 x_i + C_2 x_i^2 + \cdots + C_k x_i^k)](-x_i^2) = 0$$

$$\cdots\cdots\cdots\cdots\cdots\cdots\cdots\cdots\cdots\cdots\cdots\cdots\cdots\cdots\cdots\cdots\cdots\cdots\cdots\cdots\cdots\cdots\cdots$$

$$\frac{\partial S}{\partial C_k} = 2 \sum_{i=1}^{n} [f(x_i) - (C_0 + C_1 x_i + C_2 x_i^2 + \cdots + C_k x_i^k)](-x_i^k) = 0$$

Rearranging the above equations leads to the following linear simultaneous equations.

$$C_0 n + C_1 \sum_{i=1}^{n} x_i + C_2 \sum_{i=1}^{n} x_i^2 + \cdots + C_k \sum_{i=1}^{n} x_i^k = \sum_{i=1}^{n} f(x_i)$$

$$C_0 \sum_{i=1}^{n} x_i + C_1 \sum_{i=1}^{n} x_i^2 + C_2 \sum_{i=1}^{n} x_i^3 + \cdots + C_k \sum_{i=1}^{n} x_i^{k+1} = \sum_{i=1}^{n} x_i f(x_i)$$

$$C_0 \sum_{i=1}^{n} x_i^2 + C_1 \sum_{i=1}^{n} x_i^3 + C_2 \sum_{i=1}^{n} x_i^4 + \cdots + C_k \sum_{i=1}^{n} x_i^{k+2} = \sum_{i=1}^{n} x_i^2 f(x_i)$$

$$\cdots\cdots\cdots\cdots\cdots\cdots\cdots\cdots\cdots\cdots\cdots\cdots\cdots\cdots\cdots\cdots\cdots\cdots\cdots\cdots\cdots$$

$$C_0 \sum_{i=1}^{n} x_i^k + C_1 \sum_{i=1}^{n} x_i^{k+1} + C_2 \sum_{i=1}^{n} x_i^{k+2} + \cdots + C_k \sum_{i=1}^{n} x_i^{2k} = \sum_{i=1}^{n} x_i^k f(x_i)$$

$$\textbf{(4.8)}$$

These are $(k+1)$ normal equations in $(k+1)$ unknowns and thus can be solved for $C_0, C_1, C_2 \dots C_k$.

Example 4.6.

Fit a second-degree polynomial to the data points given in the following table.

x	1	2	3	4	5	6	7	8	9	10
y	16	30	52	82	120	166	220	282	352	430

Solution

To fit a second degree polynomial to the data points using lest square method, consider the following table.

x	y	x^2	x^3	x^4	xy	$x^2 y$
1	16	1	1	1	16	16
2	30	4	8	16	60	120
3	52	9	27	81	156	468
4	82	16	64	256	328	1312

5	120	25	125	625	600	3000
6	166	36	216	1296	996	5976
7	220	49	343	2401	1540	10780
8	282	64	512	4096	2256	18048
9	352	81	729	6561	3168	28512
10	430	100	1000	10000	4300	43000
$\sum x = 55$	$\sum y = 1750$	$\sum x^2 = 385$	$\sum x^3 = 3025$	$\sum x^4 = 25333$	$\sum xy = 13420$	$\sum x^2 y = 111232$

Therefore, the linear simultaneous equations are

$$10C_0 + 55C_1 + 385C_2 = 1750$$

$$55C_0 + 385C_1 + 3025C_2 = 13420$$

$$385C_0 + 3025C_1 + 25333C_2 = 111232$$

From which we obtain, $C_0 = 10$, $C_1 = 2$ and $C_2 = 4$. Thus the second degree polynomial is $y = 10 + 2x + 4x^2$.

LINEAR FORM OF NONLINEAR FUNCTION FOR CURVE FITTING

In many experiments, data may have other form rather than linear and polynomial forms. The nature of the curve to be fitted can be determined by inspecting the data points and considering physical phenomena. For example, the current in a dc RC circuit or voltage across the resistor varies exponentially with time. Thus, the data points must have exponential form. In this case, the assumption of an exponential function to fit the data points is more appropriate. Many nonlinear functions like as exponential function, power function, or reciprocal function, can be linearized by suitable transformations, which can be applied for linear least-squares regression. This section discusses linearization of some popular functions that arise experimentally.

Let consider following form of an exponential function to be fitted.

$$y = ae^{-bx} \tag{4.9}$$

This equation is linearized by taking logarithm of *y*:

$$\ln y = \ln a - bx$$

Setting $Y = \ln y$, $C_0 = \ln a$, $C_1 = -b$ and $X = x$ the above equation takes the following form:

$$Y = C_0 + C_1 X, \tag{4.10}$$

which is a linear form of Eq. (4.9). The coefficients C_0 and C_1 can be obtained from Eq. (4.10). Once the coefficients C_0 and C_1 are found the constants in the exponential expression can be obtained as:

$$a = e^{C_0}$$
$$b = -C_1$$

The following table shows the transformation of other nonlinear functions in the linearized form $Y = C_0 + C_1 X$.

Table 4.3. Different nonlinear functions in the form $Y = C_0 + C_1 X$.

Nonlinear function	Linearized form	Change of Variables			
$y = f(x)$	$Y = C_0 + C_1 X$	X	Y	a	b
$y = ae^{bx}$	$\ln y = \ln a + bx$	x	$\ln y$	e^{C_0}	C_1
$y = axe^{-bx}$	$\ln\left(\frac{y}{x}\right) = \ln a - bx$	x	$\ln\left(\frac{y}{x}\right)$	e^{C_0}	$-C_1$
$y = ax^b$	$\ln y = \ln a + b\ln x$	$\ln x$	$\ln y$	e^{C_0}	C_1
$y = a + \frac{b}{x}$	$y = a + b\frac{1}{x}$	$\frac{1}{x}$	y	C_0	C_1
$y = \frac{1}{a + bx}$	$\frac{1}{y} = a + bx$	x	$\frac{1}{y}$	C_0	C_1
$y = \frac{ax}{b + x}$	$\frac{1}{y} = \frac{1}{a} + \frac{b}{a}\frac{1}{x}$	$\frac{1}{x}$	$\frac{1}{y}$	$\frac{1}{C_0}$	$\frac{C_1}{C_0}$

Example 4.7.

The table below gives experimental values of x and y. Assuming that exponential function $y = ae^{bx}$ fits the following data, find the value of a and b.

x	2	4	6	8	10
y	5.98	19.84	65.88	218.72	726.17

Solution

The given function is

$$y = ae^{bx}$$

Taking logarithm on both sides

$$\ln y = \ln a + bx$$

Let $Y = \ln y$, $C_0 = \ln a$, $C_1 = b$ and $X = x$ the above equation takes the following form

$$Y = C_0 + C_1 X$$

To solve this equation using least square method, consider the following table.

$X = x$	y	$Y = \ln y$	X^2	XY
2	5.98	1.79	4	3.58
4	19.84	2.99	16	11.95
6	65.88	4.19	36	25.13
8	218.72	5.39	64	43.10
10	726.17	6.59	100	65.88
$\sum X = 30$		$\sum Y = 20.94$	$\sum X^2 = 220$	$\sum XY = 149.63$

For linear least-squares method linear simultaneous equations are

$$C_0 n + C_1 \sum_{i=1}^{n} X_i = \sum_{i=1}^{n} f(X_i)$$

$$C_0 \sum_{i=1}^{n} X_i + C_1 \sum_{i=1}^{n} X_i^2 = \sum_{i=1}^{n} X_i f(X_i)$$

Therefore,

$$5C_0 + 30C_1 = 20.94$$

$$30C_0 + 220C_1 = 149.63$$

From which we obtain, $C_0 = 0.5895$ and $C_1 = 0.6$. Therefore, $a = e^{C_0} = 1.8$ and $b = C_1 = 0.6$.

APPROXIMATION

Approximation of explicitly known functions, such as sine, cosine, and exponential, can be presented in many ways to describe their properties. However, more efficient and accurate approximations are necessary to evaluate a mathematical function on a computer. One way to compute a mathematical function on a computer is to store the functional values for a range of discrete arguments and use interpolation for nontabulated values. This method is not economical because of requiring a data table to store for each explicit function. Therefore, a suitable form of approximation is necessary. There are many forms of such approximation techniques for example, polynomial, trigonometric, and rational function. Polynomial functions are popular approximation methods. The polynomial function should have a lower degree to confirm the desired accuracy to approximate a function. Additionally, as the function will be used many times, the computational time must be a minimum.

LEAST-SQUARES APPROXIMATION

Least-squares approximations are commonly used methods for approximating an explicit function $f(x)$ on an interval [a, b]. Let consider a kth degree polynomial

$$\phi(x) = C_0 + C_1 x + C_2 x^2 + \cdots + C_k x^k \qquad \textbf{(4.11)}$$

The coefficients in Eq. (4.11) are selected to minimize the sum of the square of distances

$$S(C_0, C_1, C_2 \ldots C_k) = \int_a^b W(x)[f(x) - (C_0 + C_1 x + C_2 x^2 + \cdots + C_k x^k)]^2 \, dx$$

To have a minimum value

$$\frac{\partial S}{\partial C_i} = 0 \qquad\qquad (4.12)$$

where $i = 0, 1, 2, \ldots, k$. Assume $W(x) = 1$. The system in Eqs. (4.12) results in $k + 1$ equations in $k + 1$ unknown which can be solved for $C_0, C_1, C_2 \ldots C_k$.

$$\frac{\partial S}{\partial C_0} = 2 \int_a^b [f(x) - (C_0 + C_1 x + C_2 x^2 + \cdots + C_k x^k)](-1)\, dx = 0$$

$$\frac{\partial S}{\partial C_1} = 2 \int_a^b [f(x) - (C_0 + C_1 x + C_2 x^2 + \cdots + C_k x^k)](-x)\, dx = 0$$

$$\frac{\partial S}{\partial C_2} = 2 \int_a^b [f(x) - (C_0 + C_1 x + C_2 x^2 + \cdots + C_k x^k)](-x^2)\, dx = 0$$

$$\cdots\cdots\cdots\cdots\cdots\cdots\cdots\cdots\cdots\cdots\cdots\cdots\cdots\cdots\cdots\cdots$$

$$\frac{\partial S}{\partial C_k} = 2 \int_a^b [f(x) - (C_0 + C_1 x + C_2 x^2 + \cdots + C_k x^k)](-x^k)\, dx = 0$$

Rearranging the above equations leads to the following linear simultaneous equations

$$\int_a^b (C_0 + C_1 x + C_2 x^2 + \cdots + C_k x^k)\, dx = \int_a^b f(x)\, dx$$

$$\int_a^b (C_0 x + C_1 x^2 + C_2 x^3 + \cdots + C_k x^{k+1})\, dx = \int_a^b x f(x)\, dx$$

$$\cdots\cdots\cdots\cdots\cdots\cdots\cdots\cdots\cdots\cdots\cdots\cdots\cdots\cdots\cdots\cdots$$

$$\int_a^b (C_0 x^k + C_1 x^{k+1} + C_2 x^{k+2} + \cdots + C_k x^{2k})\, dx = \int_a^b x^k f(x)\, dx$$

Example 4.8.

Obtain fourth degree least-squares polynomial expression to approximate the function $f(x) = e^x$ on $-1 \le x \le 1$. Assume $W(x) = 1$.

Solution

Let the fourth degree polynomial is $= C_0 + C_1 x + C_2 x^2 + C_3 x^3 + C_4 x^4$. Using least-squares approximation, the following system can be found

$$C_0 \int_{-1}^{1} dx + C_1 \int_{-1}^{1} x dx + C_2 \int_{-1}^{1} x^2 dx + C_3 \int_{-1}^{1} x^3 dx + C_4 \int_{-1}^{1} x^4 dx = \int_{-1}^{1} e^x dx$$

$$C_0 \int_{-1}^{1} x dx + C_1 \int_{-1}^{1} x^2 dx + C_2 \int_{-1}^{1} x^3 dx + C_3 \int_{-1}^{1} x^4 dx + C_4 \int_{-1}^{1} x^5 dx$$
$$= \int_{-1}^{1} x e^x dx$$

$$C_0 \int_{-1}^{1} x^2 dx + C_1 \int_{-1}^{1} x^3 dx + C_2 \int_{-1}^{1} x^4 dx + C_3 \int_{-1}^{1} x^5 dx + C_4 \int_{-1}^{1} x^6 dx$$
$$= \int_{-1}^{1} x^2 e^x dx$$

$$C_0 \int_{-1}^{1} x^3 dx + C_1 \int_{-1}^{1} x^4 dx + C_2 \int_{-1}^{1} x^5 dx + C_3 \int_{-1}^{1} x^6 dx + C_4 \int_{-1}^{1} x^7 dx$$
$$= \int_{-1}^{1} x^3 e^x dx$$

$$C_0 \int_{-1}^{1} x^4 dx + C_1 \int_{-1}^{1} x^5 dx + C_2 \int_{-1}^{1} x^6 dx + C_3 \int_{-1}^{1} x^7 dx + C_4 \int_{-1}^{1} x^8 dx$$
$$= \int_{-1}^{1} x^4 e^x dx$$

which results in

$$\left[C_0 x + C_1 \frac{x^2}{2} + C_2 \frac{x^3}{3} + C_3 \frac{x^4}{4} + C_4 \frac{x^5}{5} \right]_{-1}^{1} = [e^x]_{-1}^{1}$$

$$\left[C_0\frac{x^2}{2} + C_1\frac{x^3}{2} + C_2\frac{x^4}{3} + C_3\frac{x^5}{4} + C_4\frac{x^6}{5}\right]_{-1}^{1} = [e^x(x-1)]_{-1}^{1}$$

$$\left[C_0\frac{x^3}{2} + C_1\frac{x^4}{2} + C_2\frac{x^5}{3} + C_3\frac{x^6}{4} + C_4\frac{x^7}{5}\right]_{-1}^{1} = [e^x(x^2-2x+2)]_{-1}^{1}$$

$$\left[C_0\frac{x^4}{2} + C_1\frac{x^5}{2} + C_2\frac{x^6}{3} + C_3\frac{x^7}{4} + C_4\frac{x^8}{5}\right]_{-1}^{1} = [e^x(x^3-3x^2+6x-6)]_{-1}^{1}$$

$$\left[C_0\frac{x^5}{2} + C_1\frac{x^6}{2} + C_2\frac{x^7}{3} + C_3\frac{x^8}{4} + C_4\frac{x^9}{5}\right]_{-1}^{1}$$

$$= [e^x(x^4-4x^3+12x^2-24x+24)]_{-1}^{1}$$

Therefore, the linear simultaneous equations are

$$2C_0 + 0 + 0.66667\,C_2 + 0 + 0.4C_4 = 2.3504$$

$$0 + 0.66667C_1 + 0 + 0.4C_3 + 0 = 0.73576$$

$$0.66667\,C_0 + 0 + 0.4C_2 + 0 + 0.28571C_4 = 0.87888$$

$$0 + 0.4C_1 + 0 + 0.28571C_3 + 0 = 0.44951$$

$$0.4C_0 + 0 + 0.28571\,C_2 + 0 + 0.22222C_4 = 0.55237$$

These are five equations in five unknown which results in $C_0 = 1$, $C_1 = 0.99795$, $C_2 = 0.49935$, $C_3 = 0.17614$, and $C_4 = 0.043597$. Therefore, the desired polynomial becomes

$$y = 1 + 0.99795x + 0.49935x^2 + 0.17614x^3 + 0.043597x^4.$$

On the interval [-1,1], the third-degree polynomial expression of $f(x) = e^x$ is given by

$$y = 0.99629 + 0.99795x + 0.53672x^2 + 0.17614x^3.$$

The error of the third and fourth-degree polynomial on the interval [-1,1] is shown in the following Fig. (**4.6**).

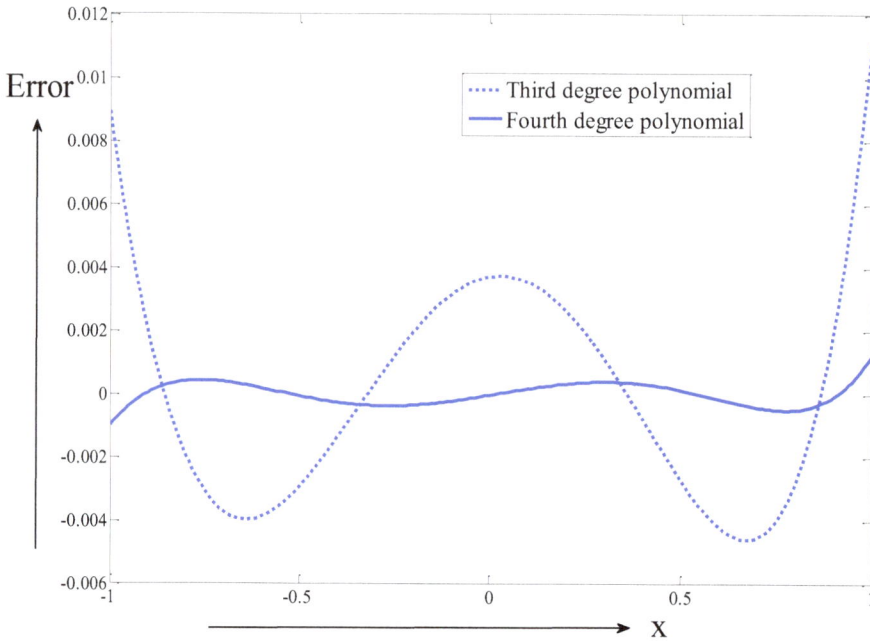

Fig. (4.6). Error in the third and fourth degree polynomial approximation of $f(x) = e^x$.

Example 4.9.

For the function $f(x) = \sin(x)$ on $0 \le x \le \pi/2$, find third degree least-squares polynomial. Assume $W(x) = 1$.

Solution

Let the third degree polynomial is $y = C_0 + C_1 x + C_2 x^2 + C_3 x^3$. Using least-squares approximation the following system can be found

$$C_0 \int_0^{\pi/4} dx + C_1 \int_0^{\pi/4} x\,dx + C_2 \int_0^{\pi/4} x^2\,dx + C_3 \int_0^{\pi/4} x^3\,dx = \int_0^{\pi/4} \sin(x)\,dx$$

$$C_0 \int_0^{\pi/4} x\,dx + C_1 \int_0^{\pi/4} x^2\,dx + C_2 \int_0^{\pi/4} x^3\,dx + C_3 \int_0^{\pi/4} x^4\,dx = \int_0^{\pi/4} x\sin(x)\,dx$$

$$C_0 \int_0^{\pi/4} x^2 dx + C_1 \int_0^{\pi/4} x^3 dx + C_2 \int_0^{\pi/4} x^4 dx + C_3 \int_0^{\pi/4} x^5 dx$$

$$= \int_0^{\pi/4} x^2 \sin(x) dx$$

$$C_0 \int_0^{\pi/4} x^3 dx + C_1 \int_0^{\pi/4} x^4 dx + C_2 \int_0^{\pi/4} x^5 dx + C_3 \int_0^{\pi/4} x^6 dx$$

$$= \int_0^{\pi/4} x^3 \sin(x) dx$$

Excluding the integration limits the normal equations are

$$C_0 x + C_1 \frac{x^2}{2} + C_2 \frac{x^3}{3} + C_3 \frac{x^4}{4} = -\cos x$$

$$C_0 \frac{x^2}{2} + C_1 \frac{x^3}{3} + C_2 \frac{x^4}{4} + C_3 \frac{x^5}{5} = -x \cos x + \sin x$$

$$C_0 \frac{x^3}{3} + C_1 \frac{x^4}{4} + C_2 \frac{x^5}{5} + C_3 \frac{x^6}{6} = -x^2 \cos x + 2x \sin x + 2 \cos x$$

$$C_0 \frac{x^4}{4} + C_1 \frac{x^5}{5} + C_2 \frac{x^6}{6} + C_3 \frac{x^7}{7} = -x^3 \cos x + 3x^2 \sin x + 6x \cos x - 6 \sin x$$

The system can be written as

$$CA = B,$$

where

$$A = \begin{bmatrix} x & \frac{x^2}{2} & \frac{x^3}{3} & \frac{x^4}{4} \\ \frac{x^2}{2} & \frac{x^3}{3} & \frac{x^4}{4} & \frac{x^5}{5} \\ \frac{x^3}{3} & \frac{x^4}{4} & \frac{x^5}{5} & \frac{x^6}{6} \\ \frac{x^4}{4} & \frac{x^5}{5} & \frac{x^6}{6} & \frac{x^7}{7} \end{bmatrix}$$

$$B = \begin{bmatrix} -\cos x \\ -x\cos x + \sin x \\ -x^2\cos x + 2x\sin x + 2\cos x \\ -x^3\cos x + 3x^2\sin x + 6x\cos x - 6\sin x \end{bmatrix}$$

Thus,

$$C = inv(A)B = \begin{bmatrix} -0.0022584 \\ 1.0272 \\ -0.069943 \\ -0.11387 \end{bmatrix}$$

Therefore, the desired polynomial becomes

$$y = -0.0022584 + 1.0272x - 0.069943x^2 - 0.11387x^3$$

On the interval $[0, \pi/2]$, the second-degree polynomial expression of $f(x) = \sin x$ is given by

$$y = -0.024325 + 1.1957x - 0.33824x^2$$

The error in the second and third-degree polynomial on the interval $[0, \pi/2]$ is shown in the following Fig. (**4.7**).

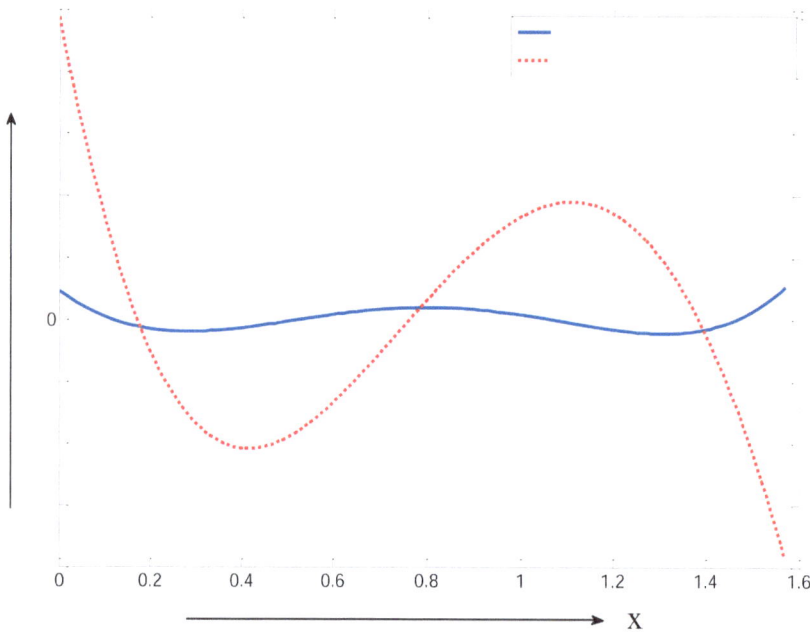

Fig. (4.7). Error in the second and third degree polynomial approximation of $f(x) = \sin x$.

CONSENT FOR PUBLICATION

Not applicable.

CONFLICT OF INTEREST

The author declares no conflict of interest, financial or otherwise.

ACKNOWLEDGEMENT

Declared none.

REFERENCES

[1] M. K, Jain, K. Rajendra, and S. R. K. Iyengar, *Numerical Methods for Scientific and Engineering Computation*. Wiley, 1985.

[2] Jaluria and Yogesh, *Computer Methods for Engineering*. Allyn and Bacon, 1988.

[3] V. Zalizniak, *Essentials of Scientific Computing : Numerical Methods for Science and Engineering*. Chichester, UK: Horwood Pub., 2008.

[4] K. E. Atkinson, *An Introduction to Numerical Analysis*. New York: Wiley, 1978.

[5] J. F. Epperson, *An Introduction to Numerical Methods and Analysis*. Hoboken: Wiley, 2013.

EXERCISE

4.1 The data table as below follows a function $f(x)=2x^2+3x+4$

x	0	2	4	6
y	5	21	53	101

Compute $y(0.15)$ and $y(3)$ using second and third order polynomial.

4.2 Consider a function $y = sin(x)$ where x is varied from 0 to π with interval $\pi/4$. Compute $y(\pi/3)$ using Lagrange polynomial. Determine required order of the polynomial to get lower error in computation.

4.3 Compute $y(2.5)$ using 2nd, 3rd and 4th order Lagrange interpolating polynomial for the following data table.

x	1	2	3	4	5
y	5	12	31	68	129

4.4 The temperature distribution in a room is given below. Find the temperature at time $t = 10$ sec.

t (sec)	2	4	6	8	12
L (m)	22.4	83.6	185.6	328.4	736.4

4.5 The following table shows position of an object moving upward from the ground level with time. Find the position of the object at time $t = 3.5$ sec.

t (sec)	1	3	5	7	9	11
h (m)	195.1	555.9	877.5	1159.9	1403.1	1607.1

4.6 The distance travelled by a car with time is as follows. Find the position of the car at time $t = 2$ and 4.5 sec.

t (sec)	1	3	5	7	9	11
d (m)	12.5	52.5	112.5	192.5	292.5	412.5

4.7 Suppose a robotic arm is moving in a circular path. Find the angular velocity of the arm at $t = 11$ sec from the following table.

t (sec)	2	4	6	8	10	12
ω (rad/sec)	20	30	40	50	60	70

4.8 The population in a city was found as below.

Year	1810	1820	1830	1840	1850	1860	1870
Population (Thousand)	32	55	69	88	95	118	139

Estimate population in the year 1815 and 1866.

4.9 The tabulated points of a function $y = f(x)$ are shown below. Find the value of $y(3.5)$ and $y(7)$.

x	1	2	3	4	5	6
y	0.2	1.6	5.4	12.8	25	43.2

4.10 Consider a function of $y(x) = \cos x$ in the range $0.1 \leq x \leq 0.5$ by increasing x with 0.1 interval. Compare the forward and backward interpolation formula for interpolation and extrapolation.

4.11 Find the best values of a and b of a function $y = ae^b$ that fit the following data table.

x	1	3	5	7	9
y	0.89	4.41	21.84	108.17	535.77

4.12 Find the constant a and b using least square method such that $y = ax^b$ fit the following data.

x	2	4	6	8	10
y	16	64	144	256	400

4.13 A nonlinear function of the form $y = \dfrac{ax}{b+x}$ fits the following table.

x	2	4	6	8	10
y	0.67	1.00	1.20	1.33	1.43

Find the best values of a and b.

4.14 For the function $y = \cos(x)$ in the interval $0 \le x \le \pi/2$, find second, third and fourth degree least-squares polynomial and compare the results.

4.15 Current I through an external resistance R in an electrical circuit where m number of electrical sources connected in parallel with voltage E is expressed by

$I = \dfrac{mE}{mR + r}$, where r is the internal resistance of the source. Assuming $E = 1.5$ V, determine the best value of m and r from the following data table

R (ohm)	1	3	5	7	9
I (A)	0.9375	0.3261	0.1974	0.1415	0.1103

4.16. The resonant wavelengths in a fibre optic sensing system for the change in the sample refractive index are given below. Fit a second and third degree polynomial to the data points.

Refractive index	1.33	1.34	1.35	1.36	1.37	1.38	1.39	1.40
Resonant wavelength (nm)	600	610	630	650	730	780	890	1070

CHAPTER 5

Introduction of Numerical Differentiation and Integration

Md. Rashidul Islam[1]*, Md. Shamim Anower[1] and Md. Mahabubur Rahman[1]

[1] Department of Electrical & Computer Engineering, Rajshahi University of Engineering & Technology, Bangladesh

Abstract: In this chapter, we study numerical differentiation and integration. At first, fundamental theories on differentiation and integration are discussed. Then, Newton's forward difference, Leibniz's notation and Lagrange's notation are presented. At the end of the chapter, various types of integration methods are discussed with the engineering problem analysis.

Keywords: Numerical differentiation, numerical integration, Newton's forward difference formula, Leibniz's notation, Lagrange's notation, Newton's notation, Euler's notation, trapezoidal rule, Simpson's 1/3 rule, Simpson's 3/8 rule, double integration, Romberg integration.

INTRODUCTION

Engineers must continuously deal with systems and processes that change, making calculus an essential tool of the engineering profession [1]. Differentiation and integration are basic mathematical operations with a wide range of applications in many areas of science and engineering [2]. It is, therefore, important to have good methods to compute and manipulate derivatives and integrals.

NUMERICAL DIFFERENTIATION

Introduction

Anything can be represented as a system, and the system can change its phase at any time. We can calculate this change analytically and numerically. Differentiation

*Corresponding author Md. Rashidul Islam: Department of Electrical & Electronic Engineering, Rajshahi University of Engineering & Technology, Bangladesh; E-mail: rashidul@eee.ruet.ac.bd

Md. Masud Rana, Wei Xu, and Youguang Guo (Eds.)

comes from change. Differentiation can be defined as the derivative, which is the instantaneous rate of change of a function with respect to one of its variables [3]. Graphical representation of the differentiation is shown in Fig. (**5.1**). Numerical differentiations are very significant in terms of Electrical Engineering and also for other fields of mathematics and physics.

The mathematical definition of the derivative is as follows:

$$\frac{\Delta y}{\Delta x} = \frac{g(x + \Delta x) - g(x)}{\Delta x} \tag{5.2.1}$$

Where y and $g(x)$ are alternative representatives for the dependent variable and x is the independent variable. If the change of x is allowed to approach zero, then the difference becomes a derivative

$$\frac{dy}{dx} = \lim_{\Delta x \to 0} \frac{\Delta y}{\Delta x} = \lim_{\Delta x \to 0} \frac{g(x + \Delta x) - g(x)}{\Delta x} \tag{5.2.2}$$

Section To Tangent

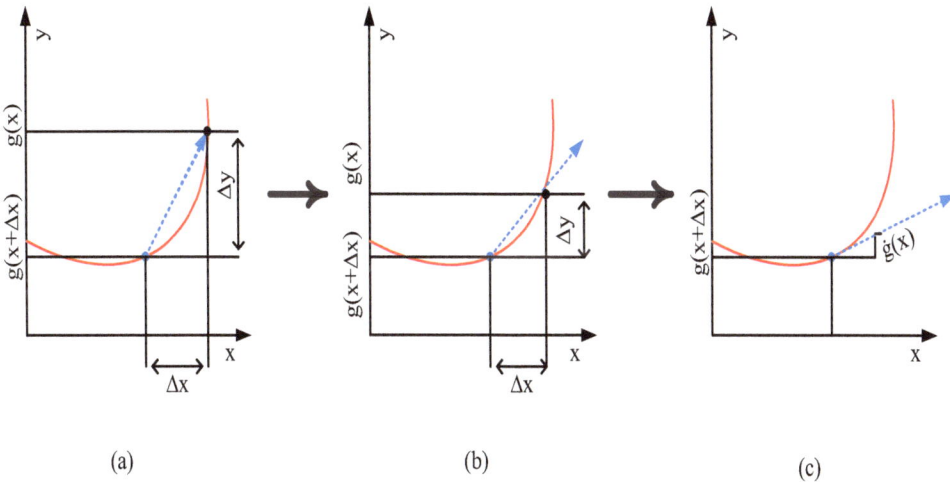

(a)   (b)   (c)

Fig. (5.1). Graphical representation of the differentiation.

As Δx approaches zero from (*a*) to (*c*), the difference approximation becomes a derivative. Here, dg/dx is the first derivative of y with respect to x evaluated at x. The first derivative is the slope of the tangent to the curve at x. The second derivative means the rate of change of the slopes of the first derivatives and can be represented as

$$\frac{d^2g}{dx^2} = \frac{d}{dx}\left(\frac{dg}{dx}\right)$$ **(5.2.3)**

Differentiation Symbols:

➢ Leibniz's notation: $\dfrac{dg}{dx}, \dfrac{d^2g}{dx^2}, \ldots\ldots, \dfrac{d^ng}{dx^n}$

➢ Newton's Notation: $\dot{g}, \ddot{g}, \dddot{g}$;

➢ Lagrange's Notation: g', g", g"', $g^{IV}, g^{V} \ldots\ldots, g^{n}$

➢ Euler's Notation: Dg, $D^2g, D^3g, \ldots\ldots, D^ng$

Why Numerical Calculus is Needed

Two problems may arise

- The function is very difficult to solve analytically, but easy numerically. Let us see an example

$$\int \left(\frac{a+x}{b-x}\right)^{\frac{1}{2}} dx = -\sqrt{a+x}\sqrt{b-x} - (a+b)\sin^{-1}\left(\frac{b-x}{a+b}\right)^{\frac{1}{2}}$$

- In many cases, experimental tabulated data may be given for integration or differentiation, easy to calculate numerically.

First Order Numerical Differentiation

If there are some tabulated values, then using the polynomial interpolating formula, a function, such as g(x), passes through an appropriate set of points. In this section, we will explain only the derivative by using Newton's forward difference formula, although other formulas can be used for derivatives.

Differentiation of Equally Spaced Data

Differentiation of equally spaced data can be obtained using the following techniques.

A. *Newton's Forward Difference Formula*

Let us consider Newton's forward difference formula

$$g = g_0 + p\Delta g_0 + \frac{p(p-1)}{2!}\Delta^2 g_0 + \frac{p(p-1)(p-2)}{3!}\Delta^3 g_0 + \dots\dots$$
$$\dots\dots + \frac{p(p-1)(p-2)\dots(p-n+1)}{n!}\Delta^n g_0 \qquad (5.2.4)$$

where

$$x = x_0 + ph \qquad (5.2.5)$$

Then

$$\frac{dg}{dx} = \frac{dg}{dp}\frac{dp}{dx} = \frac{\frac{dg}{dp}}{\frac{dx}{dp}} = \frac{1}{h}\left(\Delta g_0 + \frac{2p-1}{2}\Delta^2 g_0 + \frac{3p^2 - 6p + 2}{6}\Delta^3 g_0 + \dots\right) \qquad (5.2.6)$$

Equation **5.2.6** is used to calculate the value of $\frac{dg}{dx}$ for non-tabular values of x. For the tabular value, it takes a simpler form; if $x = x_0$ and $p = 0$, then through equation **5.2.6,** we get

$$\left[\frac{dg}{dx}\right]_{x=x_0} = \frac{1}{h}\left(\Delta g_0 - \frac{1}{2}\Delta^2 g_0 + \frac{1}{3}\Delta^3 g_0 - \frac{1}{4}\Delta^4 g_0 + ...\right) \qquad (5.2.7)$$

B. *Newton's Backward Difference Formula*

Let us consider Newton's backward difference formula

$$g = g_0 + p\nabla g_0 + \frac{p(p+1)}{2!}\nabla^2 g_0 + \frac{p(p+1)(p+2)}{3!}\nabla^3 g_0 + ...$$
$$+ \frac{p(p+1)(p+2)...(p+n-1)}{n!}\nabla^n g_0 \qquad (5.2.8)$$

and

$$\frac{dg}{dx} = \frac{dg}{dp}\frac{dp}{dx} = \frac{\frac{dg}{dp}}{\frac{dx}{dp}} = \frac{1}{h}\left(\nabla g_0 + \frac{2p+1}{2}\nabla^2 g_0 + \frac{3p^2+6p+2}{6}\nabla^3 g_0 + ...\right) \qquad (5.2.9)$$

For the tabular value, it takes a simpler form; if $x = x_0$ and $p = 0$, then through equation 5.2.9, we get

$$\left[\frac{dg}{dx}\right]_{x=x_0} = \frac{1}{h}\left(\nabla g_0 + \frac{1}{2}\nabla^2 g_0 + \frac{1}{3}\nabla^3 g_0 + \frac{1}{4}\nabla^4 g_0 + \frac{1}{5}\nabla^5 g_0 + ...\right) \qquad (5.2.10)$$

Algorithm:

1. Start
2. Input the values of dependent and independent variables
3. Input the desired point to be differentiated
4. Create Newton's difference table
5. Calculate $h = x_1 - x_0$

6. Determine the values of $\dfrac{dg}{dx}$ at the given point

7. Display the results

8. Stop

Flowchart for Newton's difference formula is shown in Fig. (**5.2**).

Flowchart:

```
          ( Start )
             │
   ┌─────────▼─────────┐
   │ Input tabulated data │
   └─────────┬─────────┘
             │
   ┌─────────▼─────────┐
   │ Input desired point x │
   └─────────┬─────────┘
             │
   ┌─────────▼─────────┐
   │   Calculate h and p  │
   └─────────┬─────────┘
             │
   ┌─────────▼─────────┐
   │ Determine differences │
   └─────────┬─────────┘
             │
   ┌─────────▼─────────┐
   │ Determine the Derivative │
   └─────────┬─────────┘
             │
   ┌─────────▼─────────┐
   │    Display Result    │
   └─────────┬─────────┘
             │
          ( End )
```

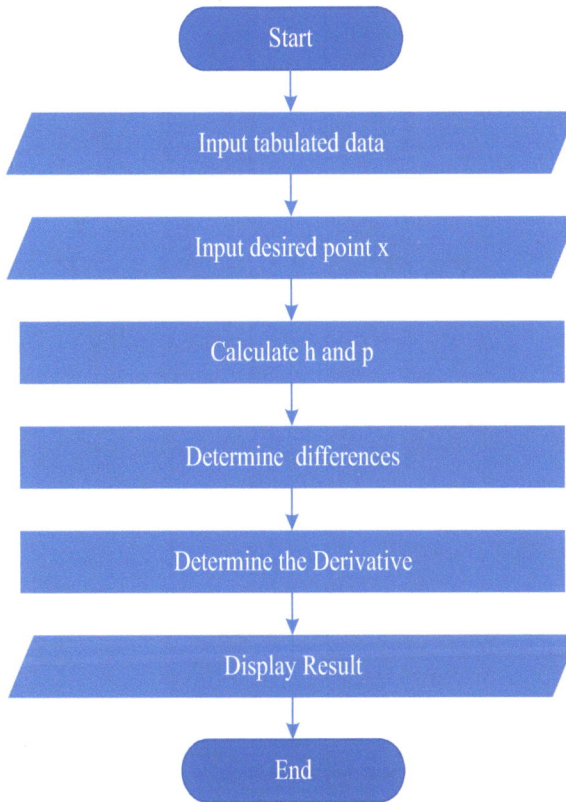

Fig (5.2). Flowchart for Newton's difference formulae.

Example 5.1: The distances traveling by an electron inside a conductor at different times (t sec) are given below:

T (sec)	0.0	0.1	0.2	0.3	0.4	0.5	0.6
X(m)	3.01	3.16	3.29	3.36	3.40	3.38	3.32

Find the velocity of the electron at time t=0.3 sec.

Solution:

T	X	Δ	Δ^2	Δ^3	Δ^4	Δ^5	Δ^6
0.0	3.01	0.15	-	-	-	-	-
-	-	-	-0.02	-	-	-	-
0.1	3.16	0.13		-0.04	-	-	-
-	-	-	-0.06	-	0.07	-	-
0.2	3.29	0.07	-	0.03	-	-0.13	-
-	-	-	-0.03	-	-0.06	-	0.24
0.3	3.36	0.04	-	-0.03	-	0.11	
-	-	-	-0.06	-	0.05	-	-
0.4	3.40	-0.02	-	0.02	-	-	-
-	-	-	-0.04	-	-	-	-
0.5	3.38	-0.06	-	-	-	-	-
-	-	-	-	-	-	-	-
0.6	3.32	-	-	-	-	-	-

Here, $T_0 = 0.3, X_0 = 3.36$ and h = 0.1, then the Eq. 5.2.7 gives

$$\text{Velocity of electron} = \left.\frac{dx}{dT}\right|_{T=0.3} = \frac{1}{0.1}\left[0.4 - \frac{1}{2}(-0.06) + \frac{1}{3}(0.02)\right]$$

$$= 0.76 \ \text{m/s}$$

Example 5.2: The switch in the following circuit is closed at t=0. After closing the switch, recorder currents for different times are tabulated as follows:

t (s)	0	1	2	3	4
i (A)	0	0.393	0.632	0.776	0.864

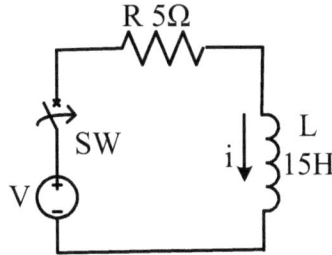

Obtain the source voltage at time t = 2s using the numerical method.
Solution:

t(s)	i (A)	Δ	Δ^2	Δ^3	Δ^4
0	0	-	-	-	-
-	-	0.393	-	-	-
1	0.393		-0.154	-	-
-	-	0.239		0.059	-
2	0.632	-	-0.095		-0.02
		0.144		0.039	-
3	0.776		-0.056	--	-
-	-	0.088	-	-	-
4	0.864	-	-	-	-

Here $t_0 = 2, i_0 = 0.632$ and h=1, then the equation 5.2.7 gives

Current passing through inductor per time= $\left. \dfrac{di}{dt} \right|_{t=2} = \left[\dfrac{1}{1} \left(0.144 - \dfrac{1}{2}(-0.056) \right) \right]$

$$=0.172 \text{ A/s}$$

Voltage across inductor = $L\dfrac{di}{dt} = 15 \times 0.172$

$$= 2.58 \text{ V}$$

Source Voltage, $V_s = V_R + V_L$

$$= I \times R + L \times \frac{di}{dt}$$

$$= 0.632 \times 5 \ + \ 15 \times 0.172$$

$$= 5.74 \text{ V}$$

Differentiation of Unequally Spaced Data

In the previous section, we discuss the differentiation for the equal interval of the independent variable. But data from experiments or field studies are often collected at unequal intervals. For this reason, such information cannot be analyzed with the techniques discussed above. To overcome the issue, second-order Lagrange interpolating polynomial formula is used.

The second-order version is:

$$g_2(x) = \frac{(x-x_1)(x-x_2)}{(x_0-x_1)(x_0-x_2)}g(x_0) + \frac{(x-x_0)(x-x_2)}{(x_1-x_0)(x_1-x_2)}g(x_1) + \frac{(x-x_0)(x-x_1)}{(x_2-x_0)(x_2-x_1)}g(x_2) \quad \textbf{(5.2.11)}$$

The second-order polynomial can be differentiated analytically to give

$$g'(x) = g(x_{i-1})\frac{2x - x_i - x_{i+1}}{(x_{i-1}-x_i)(x_{i-1}-x_{i+1})} + g(x_i)\frac{2x - x_{i-1} - x_{i+1}}{(x_i-x_{i-1})(x_i-x_{i+1})}$$
$$+ g(x_{i+1})\frac{2x - x_{i-1} - x_i}{(x_{i+1}-x_{i-1})(x_{i+1}-x_i)} \quad \textbf{(5.2.12)}$$

Where x is the value at which you want to estimate the derivative.

It has some important advantages:

1. It can be used to estimate the derivative anywhere within the range
2. Data donot need to be necessarily equally spaced
3. Accuracy is of 2nd order

Example 5.2.3: The distances traveling by an electron inside a conductor at different times (t sec) are given below:

T (s)	0.0	0.3	0.7
X(m)	3.01	3.36	3.32

Find the velocity of the electron at time t=0.3s.

Solution:

Equation 5.2.12 can be used to calculate the derivative as

$$g'(x) = 3.01\frac{2\times0.3-0.3-0.7}{(0.0-0.3)(0.0-0.7)} + 3.36\frac{2\times0.3-0.0-0.7}{(0.3-0.0)(0.3-0.7)} + 3.32\frac{2\times0.3-0.0-0.3}{(0.7-0.0)(0.7-0.3)}$$

$= 0.6238$ m/s

Numerical Differentiation of Continuous Functions

Discussion regarding numerical differentiation of continuous functions is presented below.

Forward Difference Quotient

Let $\Delta x = h$ be a small increment in x. From Taylor's theorem, we obtain,

$$g(x+h) = g(x) + hg'(x) + \frac{h^2}{2}g''(\theta) \qquad (5.2.13)$$

By rearranging the terms for $x \le \theta \le x + h$, we obtain

$$g'(x) = \frac{g(x+h) - g(x)}{h} - \frac{h}{2} g''(\theta) \qquad (5.2.14)$$

When h is very small, the equation can be written as

$$g'(x) = \frac{g(x+h) - g(x)}{h} \qquad (5.2.15)$$

In this case, the truncation error is

$$e_t(h) = -\frac{h}{2} g''(\theta) \qquad (5.2.16)$$

Here, Equation **5.2.15** is known as 1st order **Forward Difference Quotient.** It is also known as the ***two-point formula***.

The truncation error, in this case, is in order of h. The error decreases with the decrement of h.

Similarly, 1st order **Backward Difference Quotient** is

$$g'(x) = \frac{g(x) - g(x-h)}{h} \qquad (5.2.17)$$

Central Difference Quotient

1st order **Forward Difference Quotient** is a linear approximation, which results in large truncation errors in the case of higher-order functions. The error can be decreased by using quadratic approximation. It can be obtained by considering another term in Taylor's theorem, *i.e.*,

$$g(x+h) = g(x) + hg'(x) + \frac{h^2}{2!}g''(x) + \frac{h^3}{3!}g'''(\theta_1) \qquad \textbf{(5.2.18)}$$

Similarly,

$$g(x-h) = g(x) - hg'(x) + \frac{h^2}{2!}g''(x) - \frac{h^3}{3!}g'''(\theta_1) \qquad \textbf{(5.2.19)}$$

Now, subtracting Equation **5.2.18** from Equation **5.2.19**, we obtain

$$g(x+h) - g(x-h) = 2g'(x) + \frac{h^3}{3!}\left[g'''(\theta_1) + g'''(\theta_2)\right] \qquad \textbf{(5.2.20)}$$

We obtain

$$\textbf{(5.2.21)}$$

The Truncation Error in this case is

$$e_t(h) = -\frac{h^2}{12}\left[g'''(\theta_1) + g'''(\theta_2)\right] = -\frac{h^2}{6}g'''(\theta) \qquad \textbf{(5.2.22)}$$

which is of order h^2. Eq. **5.2.21** is called 1st order *Central Difference Quotient,* which is also known as the *three-point formula* and is actually the average of forward and backward difference quotients. Graphical illustration of (a) two point and (b) three point formula is shown in Fig. (**5.3**).

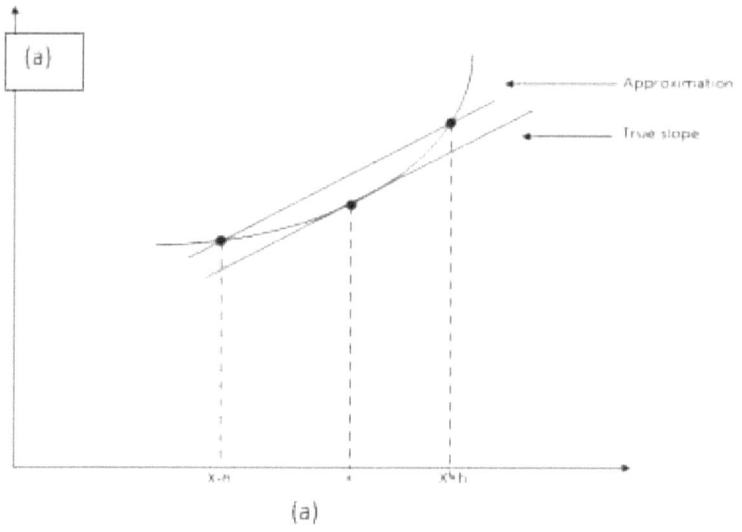

(a)

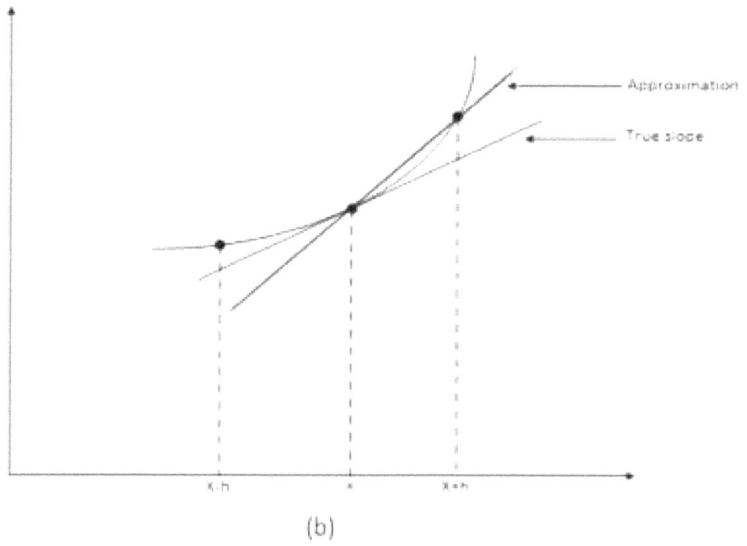

(b)

Fig. (5.3). Illustration of **(a)** two point and **(b)** three point formula.

Algorithm: Numerical Differentiation using Forward Difference formula

1. Start

2. Input the function

3. Input the desired point to be differentiated

4. Check if the function is continuous at that point

5. If the function is continuous, then move to step 6, else move to step 11

6. Input the value h

7. Determine the values of $g(x), g(x+h)$

8. Calculate the 1st derivative at point   using the following equation

$$g'(x) = \frac{g(x+h) - g(x)}{h}$$

9. Display the result

10. Stop

Algorithm: Numerical Differentiation using Central Difference formula

1. Start

2. Input the function

3. Input the desired point to be differentiated

4. Check if the function is continuous at that point

5. If the function is continuous, then move to step 6, else move to step 11

6. Input the value h

7. Determine the values of $g(x-h), g(x+h)$

8. Calculate the 1st derivative at point x using the following equation

$$g'(x) = \frac{g(x+h) - g(x-h)}{2h}$$

9. Display the result

10. Stop

Flowchart for forward and backward difference quotient is shown in Fig. (**5.4**).

Flowchart:

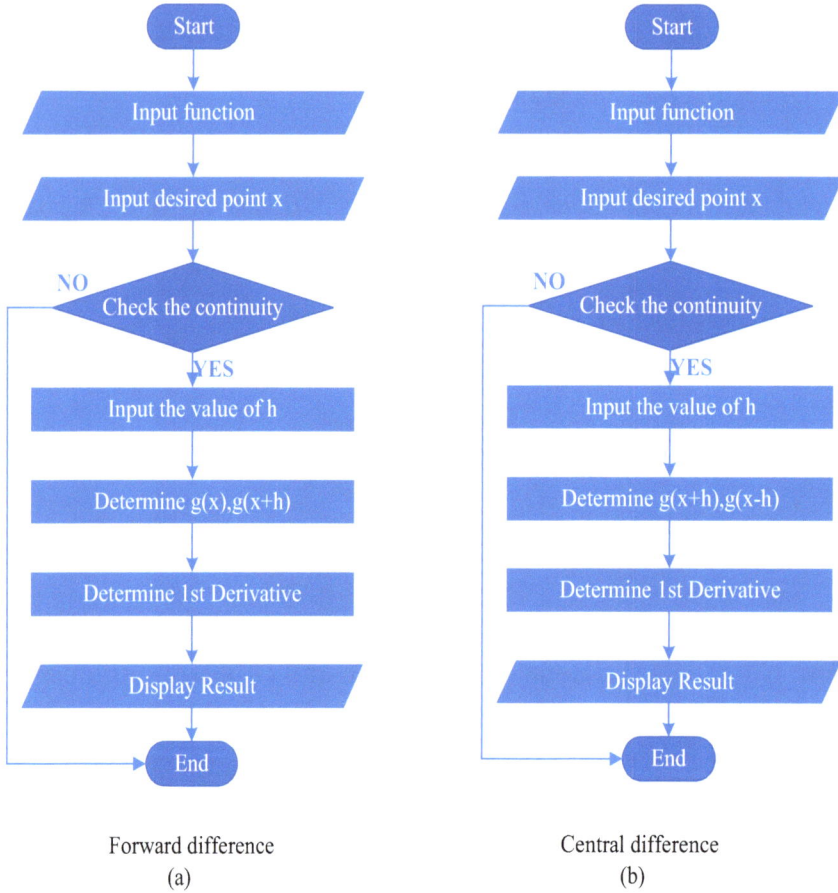

Forward difference
(a)

Central difference
(b)

Fig. (5.4). Flowchart for (a) forward and (b) backward difference quotient.

<u>Example 5.2.4:</u> The input of a differentiator circuit is given by $g(x) = x^2$. Calculate the output at $x = 1$ for $h = 0.2$ by using the forward and central difference quotient and compare the results. (The analytical solution is 2).

Solution:

Using Forward Difference Quotient

$$g'(x) = \frac{g(x+h) - g(x)}{h}$$

So,
$$g'(1) = \frac{g(1+0.2) - g(1)}{0.2} = 2.2$$

Here, error is 0.2

Using Forward Difference Quotient

So,
$$g'(1) = \frac{g(1+0.2) - g(1-0.2)}{0.4} = 2$$

Here, error is 0

Higher-Order Numerical Differentiation

As before higher-order numerical derivatives are obtained from various interpolating formulas. Different methods are discussed in subsequent sections.

Higher-Order Derivatives for Tabulated Functions

The higher-order differentiation formulae for tabulated functions can be derived from Newton's interpolation formulas. By determining the higher-order derivatives of Newton's interpolation formulas, the higher-order numerical differentiation formulas are derived. Both the forward and backward interpolation formulae can be used to determine the higher-order derivatives.

Using Newton's Forward Difference Formulae

Newton's forward difference formulae can be written as follows:

$$y = y_0 + p\Delta y_0 + \frac{p(p-1)}{2!}\Delta^2 y_0 + \frac{p(p-1)(p-2)}{3!}\Delta^3 y_0 + ... \quad (5.2.23)$$

Where,

$$x = x_0 + ph \quad \textit{(h is the step size)} \quad\quad (5.2.24)$$

Now, differentiating equation **5.2.23** twice, we obtain the 2nd order numerical differentiation formula as:

$$\frac{d^2 y}{dx^2} = \frac{1}{h^2}(\Delta^2 y_0 + \frac{6p-6}{6}\Delta^3 y_0 + \frac{12p^2 - 36p + 22}{24}\Delta^4 y_0 + ...) \quad (5.2.25)$$

This formula can be used to determine the 2nd order derivative for the value of x, which is not present in the tabulated data. For tabular values of x, if $x = x_0$, the Eq. **5.2.25** becomes:

$$\left[\frac{d^2 y}{dx^2}\right]_{x=x_0} = \frac{1}{h^2}(\Delta^2 y_0 - \Delta^3 y_0 + \frac{11}{12}\Delta^4 y_0 + ...) \quad\quad (5.2.26)$$

Algorithm:

1. Start
2. Input the tabulated data
3. Input the desired point to be differentiated
4. Calculate n = number of data
5. Calculate $h = x_1 - x_0$
6. Calculate $p = \dfrac{x - x_0}{h}$
7. Determine the values of $\Delta, \Delta^2, \Delta^3 ...... \Delta^{n-1}$
8. Calculate the 2nd derivative at point   using the following equation
$$\frac{1}{h^2}(\Delta^2 y_0 + \frac{6p-6}{6}\Delta^3 y_0 + \frac{12p^2 - 36p + 22}{24}\Delta^4 y_0 + ...)$$
9. Display the result
10. Stop

Flowchart for the 2nd order differentiation from forward difference is shown in Fig. (**5.5**).

Flowchart:

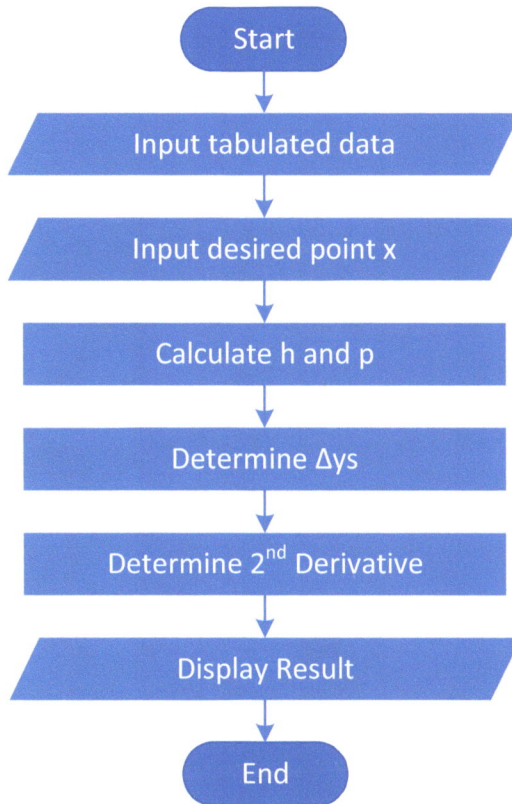

Fig. (5.5). Flowchart for the 2nd order differentiation from forward difference.

Example 5.2.6: The distances travelled by an electron inside a conductor at different times (sec) are given below:

t	0.00	0.10	0.20	0.30	0.40	0.50	0.60
x	3.01	3.16	3.29	3.36	3.40	3.38	3.32

Find out the acceleration of the electron at t=0.3s

Solution:

Here, $h = x_1 - x_0 = 0.10 - 0.00 = 0.10$

t	x	Δ	Δ^2	Δ^3	Δ^4	Δ^5	Δ^6
0.00	3.01	-	-	-	-	-	-
-	-	0.15	-	-	-	-	-
0.10	3.16	-	-0.02	-	-	-	-
-	-	0.13	-	-0.04	-	-	-
0.20	3.29	-	-0.06	-	0.07	-	-
-	-	0.07	-	0.03	-	-0.13	-
0.30	3.36	-	-0.03	-	-0.06	-	0.24
-	-	0.04	-	-0.03	-	0.11	-
0.40	3.40	-	-0.06	-	0.05	-	-
-	-	-0.02	-	0.02	-	-	-
0.50	3.38	-	-0.04	-	-	-	-
-	-	-0.06	-	-	-	-	-
0.60	3.32	-	-	-	-	-	-

The 2nd derivative of a tabulated function at x=x$_0$ is given by

$$\left[\frac{d^2 y}{dx^2}\right]_{x=x_0} = \frac{1}{h^2}\left(\Delta^2 y_0 - \Delta^3 y_0 + \frac{11}{12}\Delta^4 y_0 + ...\right)$$

The acceleration of the electron at t=0.3s,

$$\left.\frac{d^2x}{dt^2}\right|_{t=0.3} = \frac{1}{(0.1)^2}\left[(-0.06)-(0.02)\right]$$

$$= -8 \text{ ms}^{-2}$$

Ans. -8 ms^{-2}

B. *Using Newton's Backward Difference Formulae*

Newton's backward difference formulae can be written as follows:

$$y = y_n + p\nabla y_n + \frac{p(p+1)}{2!}\nabla^2 y_n + \frac{p(p+1)(p+2)}{3!}\nabla^3 y_n + ... \quad (5.2.27)$$

where,

$$x = x_n + ph \text{ (h is the step size)} \quad (5.2.28)$$

Now differentiating Eq. **5.2.27** twice, we obtain the 2nd order numerical differentiation formula:

$$\frac{d^2y}{dx^2} = \frac{1}{h^2}(\nabla^2 y_n + \frac{6p+6}{6}\nabla^3 y_n + \frac{12p^2+36p+22}{24}\nabla^4 y_n + ...) \quad (5.2.29)$$

This formula can be used to determine the 2nd order derivative for the value of x, which is not present in the tabulated data. For tabular values of *x*, if $x=x_0$, the Eq. 5.2.29 becomes

$$\left[\frac{d^2y}{dx^2}\right]_{x=x_n} = \frac{1}{h^2}(\nabla^2 y_n + \nabla^3 y_n + \frac{11}{12}\nabla^4 y_n + \frac{5}{6}\nabla^5 y_n ....) \quad (5.2.30)$$

Algorithm:

1. Start
2. Input the tabulated data
3. Input the desired point to be differentiated
4. Calculate n = number of data
5. Calculate $h = x_1 - x_0$
6. Calculate $p = \dfrac{x - x_n}{h}$
7. Determine the values of $\nabla, \nabla^2, \nabla^3 \ldots \ldots \nabla^{n-1}$
8. Calculate the 2nd derivative at point   using the following equation

$$\frac{1}{h^2}(\nabla^2 y_n + \frac{6p+6}{6}\nabla^3 y_n + \frac{12p^2 + 36p + 22}{24}\nabla^4 y_n + \ldots.)$$

9. Display the result
10. Stop

Flowchart for the 2nd order differentiation from backward difference is Fig. **(5.6).**

Flowchart:

```
                    ( Start )
                        |
                        v
         / Input tabulated data /
                        |
                        v
          / Input desired point x /
                        |
                        v
            [ Calculate h and p ]
                        |
                        v
             [ Determine ∇ys ]
                        |
                        v
         [ Determine 2^nd Derivative ]
                        |
                        v
           / Display Result /
                        |
                        v
                    ( End )
```

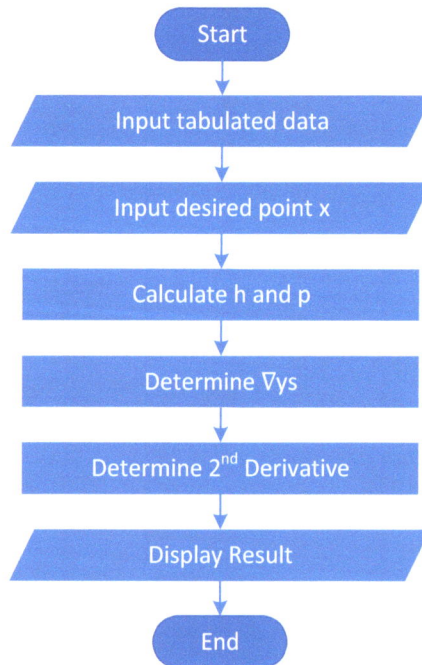

Fig. (5.6). Flowchart for the 2nd order differentiation from backward difference.

Example 5.2.7: The distances travelled by an electron inside a conductor at different times (s) are given below:

t	0.00	0.10	0.20	0.30	0.40	0.50	0.60
x	3.01	3.16	3.29	3.36	3.40	3.38	3.32

Find out the acceleration of the electron at t=0.4s

Solution:

Here, $h = x_1 - x_0 = 0.10 - 0.00 = 0.10$

t	x	∇	∇^2	∇^3	∇^4	∇^5	∇^6
0.00	3.01	-	-	-	-	-	-
-	-	0.15	-	-	-	-	-
0.10	3.16	-	-0.02	-	-	-	-
-	-	0.13	-	-0.04	-	-	-
0.20	3.29	-	-0.06	-	0.07	-	-
-	-	0.07	-	0.03	-	-0.13	-
0.30	3.36	-	-0.03	-	-0.06	-	0.24
-	-	0.04	-	-0.03	-	0.11	-
0.40	3.40	-	-0.06	-	0.05	-	-
-	-	-0.02	-	0.02	-	-	-

0.50	3.38		-0.04				
		-0.06					
0.60	3.32						

The 2nd derivative of a tabulated function at x=x$_0$ is given by:

$$\left[\frac{d^2 y}{dx^2}\right]_{x=x_n} = \frac{1}{h^2}(\nabla^2 y_n + \nabla^3 y_n + \frac{11}{12}\nabla^4 y_n + \frac{5}{6}\nabla^5 y_n....)$$

The acceleration of the electron at t=0.4s,

$$\left.\frac{d^2 x}{dt^2}\right|_{t=0.4} = \frac{1}{(0.1)^2}\left[(-0.03) + 0.03 + 0.07\right]$$

$$= 7\text{ms}^{-2}$$

Higher Order Derivatives for Continuous Functions

Higher-order derivatives for continuous functions can be obtained using Newton-divided difference formulae. Here, the central divided difference formula is used to derive the higher-order derivative. The forward divided difference method and backward divided difference method can also be used similarly.

Let $g(x)$ be the function and $g''(x)$ be the 2nd derivative. From Taylor's expansion, we obtain

$$g(x + h) = g(x) + hg'(x) + \frac{h^2}{2!}g''(x) + \frac{h^3}{3!}g'''(x) + T_1 \qquad \textbf{(5.2.31)}$$

And

$$g(x - h) = g(x) - hg'(x) + \frac{h^2}{2!}g''(x) - \frac{h^3}{3!}g'''(x) + T_2 \qquad \textbf{(5.2.32)}$$

Now adding Eq. **5.2.31** and Eq. **5.2.32,** we obtain

$$g(x+h) + g(x-h) = 2g(x) + h^2 g''(x) + T_1 + T_2 \qquad \textbf{(5.2.33)}$$

Thus,

$$g''(x) = \frac{g(x+h) - 2g(x) + g(x-h)}{h^2} - \frac{T_1 + T_2}{h^2} \qquad \textbf{(5.2.34)}$$

So, the approximation for the 2nd derivative using *Central Divided Difference Formula* is as follows

$$g''(x) = \frac{g(x+h) - 2g(x) + g(x-h)}{h^2} \text{ (Central)} \qquad \textbf{(5.2.35)}$$

The truncation error in this case is

$$e_{central}(h) = -\frac{T_1 + T_2}{h^2} = -\frac{1}{h^2}\frac{h^4}{4!}\left[g^{(4)}(\phi_1) + g^{(4)}(\phi_2)\right]$$

$$e_{central}(h) = -\frac{h^2}{12}g^{(4)}(\phi) \qquad \textbf{(5.2.36)}$$

In this method, the error we obtain is in the order of h^2

Similarly, the approximation for the 2nd derivative using *Forward Divided Difference Formula* is

$$g''(x) = \frac{2g(x) - 5g(x+h) + 4g(x+2h) - g(x+3h)}{h^2} \text{ (Forward)} \qquad \textbf{(5.2.37)}$$

The truncation error for this approximation is

$$e_{forward} = \frac{11h^2}{12} g^{(4)}(\phi) \tag{5.2.38}$$

Again, the approximation for the 2nd derivative using *Backward Divided Difference Formula* is

$$g''(x) = \frac{2g(x) - 5g(x-h) + 4g(x-2h) - g(x-3h)}{h^2} \text{ (Backward)} \tag{5.2.39}$$

The truncation error for this approximation is

$$e_{forward} = \frac{11h^2}{12} g^{(4)}(\phi) \tag{5.2.40}$$

Algorithm: (For Central Difference)

1. Start
2. Input the function
3. Input the desired point to be differentiated
4. Check if the function is continuous at that point
5. If the function is continuous, then move to step 6, else move to step 11
6. Calculate n = number of data
7. Calculate $h = x_1 - x_0$
8. Determine the values of $g(x), g(x-h), g(x+h)$
9. Calculate the 2nd derivative at point   using the following equation
   $$\frac{g(x+h) - 2g(x) + g(x-h)}{h^2}$$
10. Display the result
11. Stop

Algorithm: (For Forward Difference)

1. Start
2. Input the function
3. Input the desired point to be differentiated

4. Check if the function is continuous at that point

5. If the function is continuous, then move to step 6, else move to step 11

6. Calculate n = number of data

7. Calculate $h = x_1 - x_0$

8. Determine the values of $g(x), g(x+h), g(x+2h), g(x+3h)$

9. Calculate the 2nd derivative at point   using the following equation

$$\frac{2g(x) - 5g(x+h) + 4g(x+2h) - g(x+3h)}{h^2}$$

10. Display the result

11. Stop

Algorithm: (For Backward Difference)

1. Start

2. Input the function

3. Input the desired point to be differentiated

4. Check if the function is continuous at that point

5. If the function is continuous, then move to step 6, else move to step 11

6. Calculate n = number of data

7. Calculate $h = x_1 - x_0$

8. Determine the values of $g(x), g(x-h), g(x-2h), g(x-3h)$

9. Calculate the 2nd derivative at point   using the following equation

$$\frac{2g(x) - 5g(x-h) + 4g(x-2h) - g(x-3h)}{h^2}$$

10. Display the result

11. Stop

Flowchart for the 2nd order differentiation for continuous function is shown in Fig. (**5.7**).

Flowchart

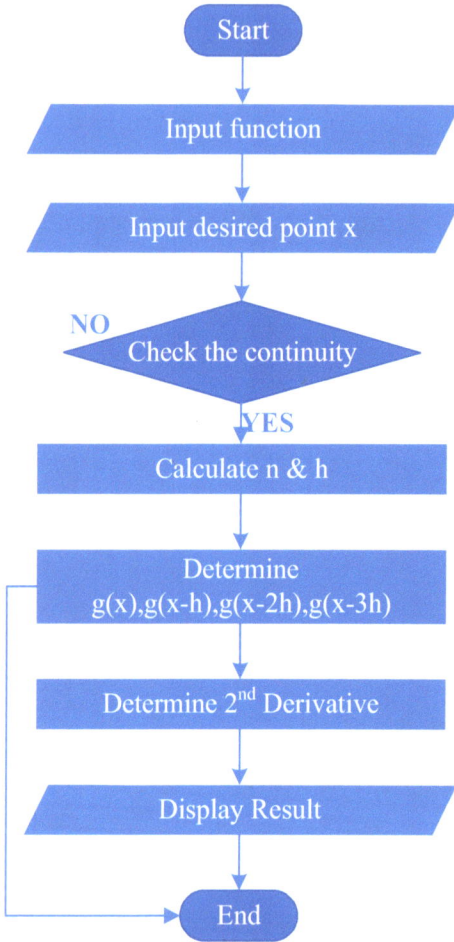

Backward Difference   **Forward Difference**

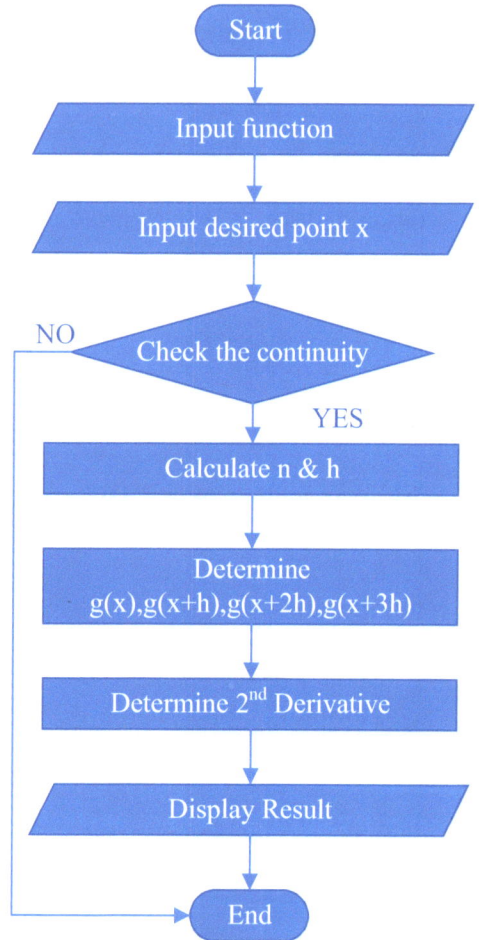

Fig. (5.7). Flowchart for the 2nd order differentiation for continuous function.

Flowchart for the 2nd order differentiation with central difference for continuous function is shown in Fig. (**5.8**).

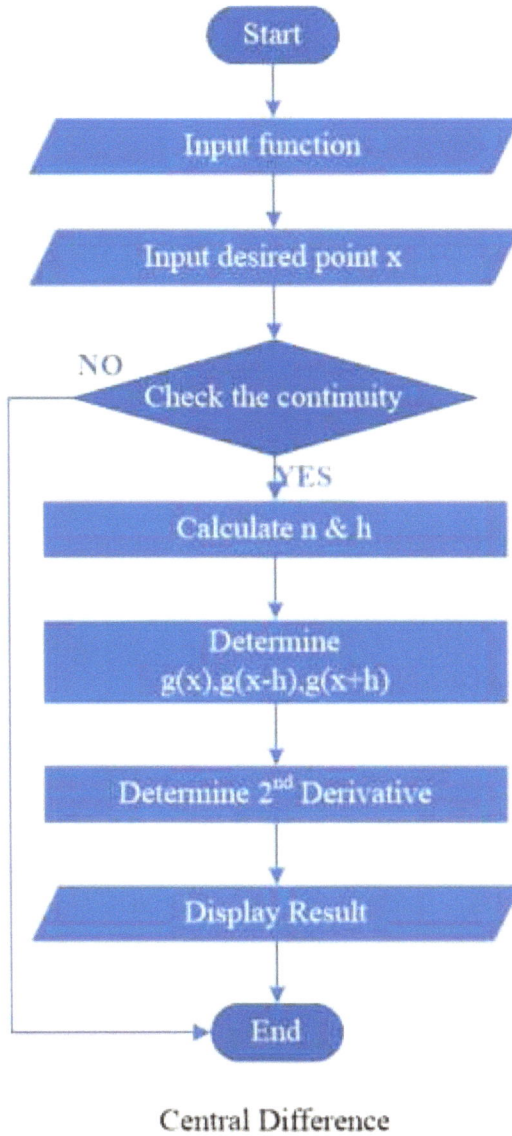

Central Difference

Fig. (5.8). Flowchart for 2nd order differentiation for continuous function.

Example 5.2.8: The input of a double differentiator circuit is given by $g(x) = x^2$. Calculate the output at $x = 1$ for $h = 0.2$ by using the forward and central difference quotient and compare the results.

Solution:

Using forward divided difference approximation,

$$g''(x)\big|_{x=1} = \frac{2g(x) - 5g(x+h) + 4g(x+2h) - g(x+3h)}{h^2}$$

$$= \frac{2(1) - 5(1.44) + 4(1.96) - 2.56}{0.2^2} = 2 \textbf{ (Ans.)}$$

Using the central divided difference approximation,

$$g''(x)\big|_{x=1} = \frac{g(x+h) - 2g(x) + g(x-h)}{h^2}$$

$$= \frac{1.44 - 2(1) + 0.64}{0.2^2} = 2 \textbf{ (Ans.)}$$

Using backward divided difference approximation,

$$g''(x)\big|_{x=1} = \frac{2g(x) - 5g(x-h) + 4g(x-2h) - g(x-3h)}{h^2}$$

$$= \frac{2(1) - 5(0.64) + 4(0.36) - 0.16}{0.2^2} = 2 \textbf{ (Ans.)}$$

NUMERICAL INTEGRATION

Introduction

Integration of a function $y = f(x)$, being well known, is nothing but the total area covered by the function, i.e., the area shown by the graphical representation of that function. Integral is the term or function to be integrated and was solved "analytically" or "directly" till now where the function is known explicitly. In the field of mathematical calculation, most of the work is done by computer programming [4]. The analytical solutions for integration are exact, but they are not very user friendly because they consume a lot of time depending on the difficulty of the function. Besides, these solutions cannot be used in computer programming as direct methods and that is where numerical integration methods come into play.

Numerical integration, as the name suggests, solves everything in some steps. In integrating numerically, the function may not be given, which was required strictly in analytical solutions. Numerical integration doesnot require the function to be known explicitly. A set of tabulated data that represents or may represent the function is enough to integrate numerically.

To integrate a function or a set of given data like,

x	$y = f(x)$
x_0	y_0
x_1	y_1
x_2	y_0
.	.
.	.
x_{n-1}	y_{n-1}
x_n	y_n

one needs to have the knowledge of interpolation. Interpolation is the process of finding the appropriate function with minimum error from a given tabulated data. The analytical representation of the function that is to be integrated is

$$I = \int_{a}^{b} f(x)dx \qquad (5.3.1)$$

Anyway, numerical integration uses Eq. **5.3.1** to find the integral of th e given tabulated data, where $f(x)$ is the function evaluated from the tabulated data using interpolation. In this chapter, numerical integration will be performed by using Newton's forward difference formula. The formula is

$$y = f(x) = y_0 + r \quad y_0 + \frac{r(r-1)}{2!}\Delta^2 y_0 + \frac{r(r-1)(r-2)}{3!}\Delta^3 y_0 + \ldots\ldots \qquad (5.3.2)$$

where,

$$r = (x - x_0)/(x_2 - x_1) \qquad (5.3.3)$$

x = The point where we want to find the functional value

y = Values got from the difference table of interpolation

Now, the limit of the integration, from the table above, is from x_0 to x_n. So, integrating Eq. **5.3.2** with respect to x, we get

$$I = \int_{x_0}^{x_n} y\,dx \qquad (5.3.4)$$

Or, $I = \int_{x_0}^{x_n} \left[y_0 + r \quad y_0 + \dfrac{r(r-1)}{2!} \Delta^2 y_0 + \dfrac{r(r-1)(r-2)}{3!} \Delta^3 y_0 + \ldots\ldots\ldots\ldots \right] dx \qquad (5.3.5)$

Now, if we denote the distance of two successive x values by h (the tabulated data will be given in uniform difference), then

$$x = x_0 + rh \quad and \quad dx = h\,dr \quad and \quad h = x_n - x_{n-1}$$

Or, $I = h \int_{0}^{n} \left[y_0 + r \quad y_0 + \dfrac{r(r-1)}{2!} \Delta^2 y_0 + \dfrac{r(r-1)(r-2)}{3!} \Delta^3 y_0 + \ldots \right] dr \qquad (5.3.6)$

Or, $I = nh \left[y_0 + \dfrac{n}{2}\Delta y_0 + \dfrac{n(2n-3)}{12} \Delta^2 y_0 + \dfrac{n(n-2)^2}{24} \Delta^3 y_0 + \ldots \right] \qquad (5.3.7)$

Eq. **5.3.7** is called "The general formula for numerical integration" and can be used for any kind of given function or tabulated data.

The Trapezoidal Method

A trapezoid consists of two parallel sides and two random sides. In numerical integration, the function graph is divided into n divisions, each creating a parallel side with the vertical axis. Thus, multiple trapezoids are formed in the graphical representation. By dividing the function, we put $n = 1$, and the method we get is known as the Trapezoidal Method. Related figures for better understanding are given below Fig. (**5.9**).

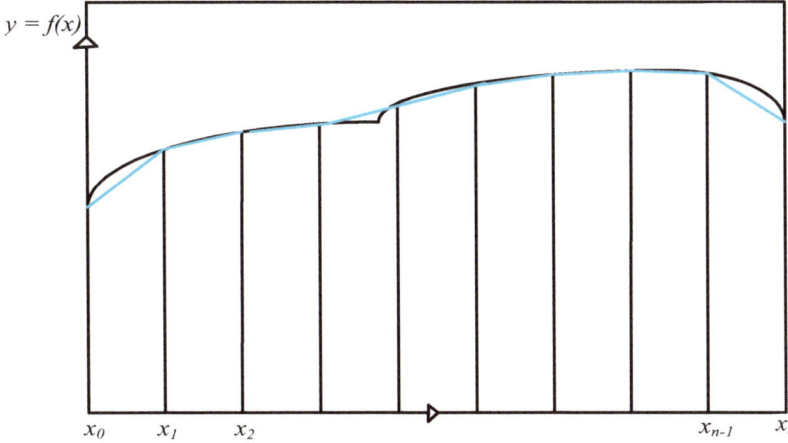

Fig. (5.9). Trapezoidal Method.

From the figure above, it can be seen that there is a number of trapezoids formed in the graph. Every trapezoid represents the area of an individual division that we made to the function. The mathematical explanation is given below. Putting $n = 1$ in Eq. 5.3.7, we get

$$\int_{x_0}^{x_1} ydx \ h(y_0 + \frac{1}{2}\Delta y_0) = h[y_0 + \frac{1}{2}(y_1 - y_0)] = \frac{h}{2}(y_0 + y_1) \qquad (5.3.8)$$

Similarly, for the next interval $[x_1, x_2]$, we have

$$\int_{x_1}^{x_2} ydx \ \frac{h}{2}(y_1 + y_2) \qquad (5.3.9)$$

Also,

$$\int_{x_{n-1}}^{x_n} y\,dx = \frac{h}{2}(y_{n-1} + y_n) \qquad (5.3.10)$$

Adding these equations, we get the formula for the trapezoidal method

$$\int_{x_0}^{x_n} y\,dx = \frac{h}{2}[y_0 + 2(y_1 + y_2 + \ldots\ldots\ldots + y_{n-1}) + y_n] \qquad (5.3.11)$$

The general steps for solving any integral using this method can be summarized as follows:

1.If the function is given, choose a value of h to divide the distance between the upper limit and lower limit uniformly.
2.Make a table as given before, from the lower limit, by h distance, to the upper limit, as the value of x. In the y column, put the functional value for each x.
3.The first y value is y_0, and the last one is y_n
4.Use Eq. 5.3.11 to solve the integration.

Algorithm:

1. Start
2. Function definition
3. Defining the x and y data along with h
4. $i = (x_n - x_0)/h$
5. while $i < n$
6. $y(i) = f(x(i))$
7. $S_1 = y_0 + y_n$ and $S_2 = 0$
8. for $i < n$
9. $S_2 = S_2 + 2*y(i)$
10. end
11. Result $= (h/2)*(S_2 + S_1)$
Stop

This algorithm can be used to program a code to solve problems by the trapezoidal method.

The flow chart for the trapezoidal method is a successive operation based on the algorithm shown above (as shown in Fig. (**5.10**)):

Flowchart:

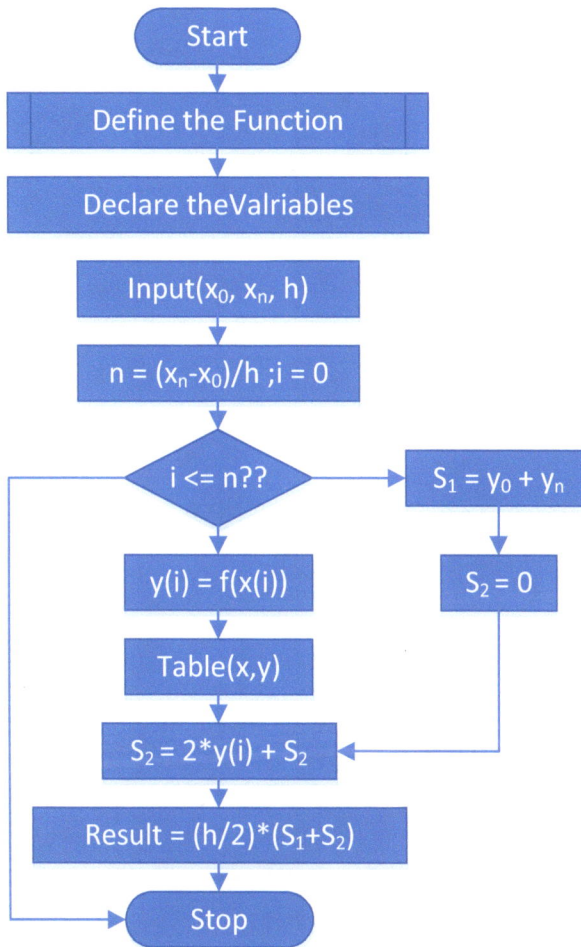

Fig. (**5.10**). Trapezoidal Method Flow Chart.

Some examples regarding this method can be illustrated in the following applications:

Example 5.3.1: Using the trapezoidal method, determine the voltage across the capacitor by assuming $v(0)$ = 0 V from t = 0 to t = 1 second for the following circuit.

    (a) Time step 0.5s   (b) Time step 0.1s

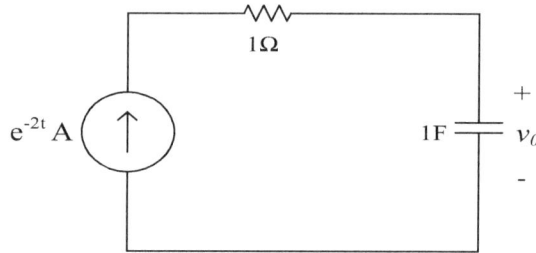

Solution:

We will follow the steps described above to solve the problem

1. If the function is given, choose a value of h to divide the distance between the upper limit and lower limit uniformly.

Here, the function to be integrated is not given. Find the function from the circuit given in the figure.

The equation for the voltage is

$$v_0 = \frac{1}{C} \int_0^1 i \, dt$$

Or, $v_0 = \int_0^1 e^{-2t} \, dt$

The analytical or exact solution is v_0 = 0.4323323584 V.

2. Make a table as given before, from the lower limit, by h distance, to the upper limit, as the value of t. In the v_0 column, put the functional value for each t.

Let time step $h = 0.5$s; the table is given by

t	v_0
0	1
0.5	0.36788
1	0.13534

The trapezoidal formula gives us

$$\int_{x_0}^{x_n} y\,dx = \frac{h}{2}\left[y_0 + 2(y_1 + y_2 + \ldots\ldots + y_{n-1}) + y_n\right]$$

So, $v_0 = \int_0^1 e^{-2t}dt = \frac{0.5}{2}(1 + 2*0.36788 + 0.13534)$

$$v_0 = 0.467775 \text{ volts}$$

Let time step $= 0.1$ seconds, the table becomes

t	v_0
0	1
0.1	0.81873
0.2	0.67032
0.3	0.54881
0.4	0.44933
0.5	0.36788
0.6	0.30119
0.7	0.2466

0.8	0.2019
0.9	0.1653
1	0.13534

The trapezoidal method gives us, using the same formula, $v_0 = 0.4337725$ volts

Example 5.3.2: The outflow concentration of a reactor is measured at a number of instants over 24 hours' period

Time, t(h)	0	4	8	12	16	20	24
C(t)(mg/L)	1	2	2.3	4	5.5	3	1.2

The flow rate of the outflow in m^3 / s can be computed with the following equation:

$$Q(t) = 20t + 10\sin\left\{\frac{2\pi}{24}(t - 10)\right\}$$

Use the numerical integration to determine the flow weighted average concentration leaving the reactor over the 24 hours period using

$$C = \frac{\int_0^t Q(t)C(t)dt}{\int_0^t Q(t)dt}$$

Solution:

Let the time step $h = 4$ hours. We get the table of functional values as below

t(hr)	C(t)(mg/L)	Q(t) (m^3/s)	Q(t)C(t)
0	1	19.54	19.54
4	2	19.72	39.44
8	2.3	19.90	45.77
12	4	20.09	80.36
16	5.5	20.27	111.485
20	3	20.45	61.35
24	1.2	20.63	24.756

By using the trapezoidal method, we have

$$\int_0^{24} Q(t)C(t)dt = \frac{h}{2}\left[y_0 + 2(y_1 + y_2 + \ldots + y_5) + y_6\right]$$

$$= 1442.212$$

$$\int_0^{24} Q(t)dt = \frac{h}{2}\left[y_0 + 2(y_1 + y_2 + \ldots + y_5) + y_6\right]$$

$$= 482.06$$

$$\text{So, } C = \frac{1442.212}{482.06} = 2.9917 \approx 3\,\text{mg/L}$$

Errors in Trapezoidal Method

All the methods used in numerical analysis are not direct or exact methods. They contain some errors after getting the result. Ideally, it is not possible to get 100% correct answers using numerical analysis, but the limit of error can be set with computer programming, i.e., we can set the tolerance limit to as small as we want the error to be. But it would not be zero anyway. The error in the trapezoidal method can be found as follows:

Let $y = f(x)$ be continuous in the interval $[x_0, x_n]$. Then, we get Taylor series around $x = x_0$

$$\int_{x_0}^{x_1} y\,dx = \int_{x_0}^{x_1} [y_0 + (x - x_0)y_0' + \frac{(x - x_0)^2}{2!} y_0'' + \ldots] \qquad (5.3.12)$$

Where $x - x_0 = h$ for the interval $[x_0, x_1]$, i.e., for $n = 1$

Also,

$$\frac{h}{2}(y_0 + y_1) = \frac{h}{2}(y_0 + y_0 + hy_0' + \frac{h^2}{2!} y_0'' + \frac{h^3}{3!} y_0''' + \ldots) \qquad (5.3.13)$$

$$= hy_0 + \frac{h^2}{2} y_0' + \frac{h^3}{4} y_0'' + \frac{h^4}{12} y_0''' + \ldots \qquad (5.3.14)$$

As we considered the integral from x_0 to x_1 only, the error is also considered for this particular range, and that is given by

$$\int_{x_0}^{x_1} y\,dx - \frac{h}{2}(y_0 + y_1) = -\frac{1}{12} h^3 y_0'' + \ldots \qquad (5.3.15)$$

Preceding in a similar way, we get the error for every interval, i.e., for $[x_1, x_2]$, $[x_2, x_3]\ldots\ldots[x_{n-1}, x_n]$, respectively. The total error for the trapezoidal method is

$$E = -\frac{1}{12} h^3 (y_0'' + y_1'' + y_2'' + \ldots\ldots + y_{n-1}'') \qquad (5.3.16)$$

In the previous example, the error is calculated as:

$$\text{If } h = 0.5 \text{ , } E = \frac{0.467775 - 0.43233235\,84}{0.4323323584} \times 100\% = 8.12\%$$

$$\text{If } h = 0.1 \text{ , } E = \frac{0.4337725 - 0.43233235\,84}{0.4323323584} \times 100\% = 0.333\%$$

We can see that by reducing the value of h the error reduces significantly. Graphically, if we divide the function into very small parts, the answer will be more accurate. The divided parts in the graph are the step size h in the mathematical derivation, which satisfies the theory. If $h \to 0$, *i.e.*, the division of the graph, $n \to \infty$, the error reduces to zero.

5.3.3 Simpson's 1/3 Method

In the case of Simpson's 1/3 method, it is required to obtain more accurate results from the trapezoidal method. Thus, if the interpolating polynomial to connect the tabulated data is of second-order (first order in the case of trapezoidal), the resulting error is reduced greatly, as shown in Fig **(5.11)** and Fig **(5.12)** for an explicit function $f(x) = 3x^3 - 2x^2 - 5$.

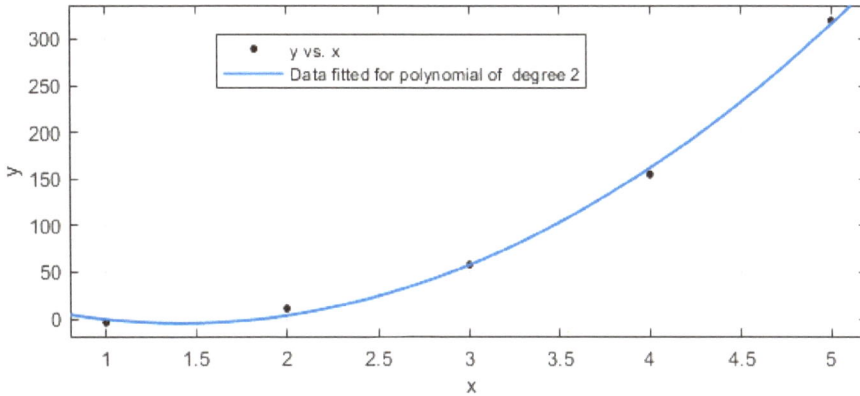

Fig. **(5.11)**. Error reduced having used second-degree polynomial.

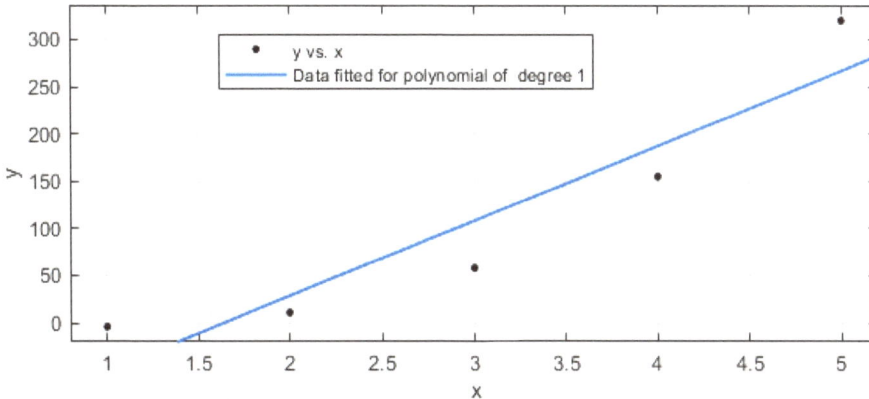

Fig. **(5.12)**. Error due to first-degree polynomial.

Hence, for the interpolating function having a second degree of polynomial, the error is reduced greatly.

So, in that case, putting n = 2, the basic numerical integration formula becomes,

$$\int_{x_0}^{x_2} y\,dx = 2h\left(y_0 + \Delta y_0 + \frac{1}{6}\Delta^2 y_0\right)$$

$$= \frac{h}{3}(y_0 + 4y_1 + y_2)$$

(5.3.17)

Similarly,

$$\int_{x_2}^{x_4} y\,dx = \frac{h}{3}(y_2 + 4y_3 + y_4)$$

.

.

.

$$\int_{x_{n-2}}^{x_n} y\,dx = \frac{h}{3}(y_{n-2} + 4y_{n-1} + y_n)$$

(5.3.18)

Finally, adding all the integrals,

$$\int_{x_0}^{x_n} y\,dx = \frac{h}{3}(y_0 + 4(y_1 + y_3 + y_5 + \ldots + y_{n-1}) + 2(y_2 + y_4 + y_6 + \ldots + y_{n-2}) + y_n)$$ (5.3.19)

Simpson's 1/3 equation is required for numerical integration.

Errors in The Simpson's 1/3 Method

As calculated in the trapezoidal methods, it can be obtained from algebraic manipulation that the error of the Simpson's 1/3 Method has a truncation error of

$$E_t = -\frac{1}{90}h^5 f^4(\xi)$$ (5.3.20)

where ξ is the point having the largest error of the fourth order derivatives. But

$h = \frac{q-p}{2}$. So, equation (2) becomes

$$E_t = -\frac{q-p}{180}h^4 f^4(\xi)$$ (5.3.21)

Hence, it has a truncation error of order four, i.e., $O(h^4)$.

Algorithm:

1. Start

2. Get the input function f(x);

3. Read the upper limit b, lower
   limit a & number of subintervals n

4. Compute h $= \frac{b-a}{n}$

5. Sum= [f(a) − f(a + nh)]

6. for i=0 to n-1

7. do compute sum=sum+4 * f(a + ih) + 2 * f(a + (i + 1)h); end

8. Integration I $=$ sum $* \frac{h}{3}$

9. Display result I

10. Stop

The flow chart of Simpson's 1/3 method is shown in Fig. **(5.13)**. This algorithm can be used to program Simpson's 1/3 Method using any kind of programming software or languages such as MATLAB, CODEBLOCKS, *etc.* Numerical methods are all about calculations, which can greatly increase one's time to solve a particular method. That is the reason why each and every method of numerical analysis should be computer programmed in order to reduce the time taken by the hand on calculations.

Flowchart:

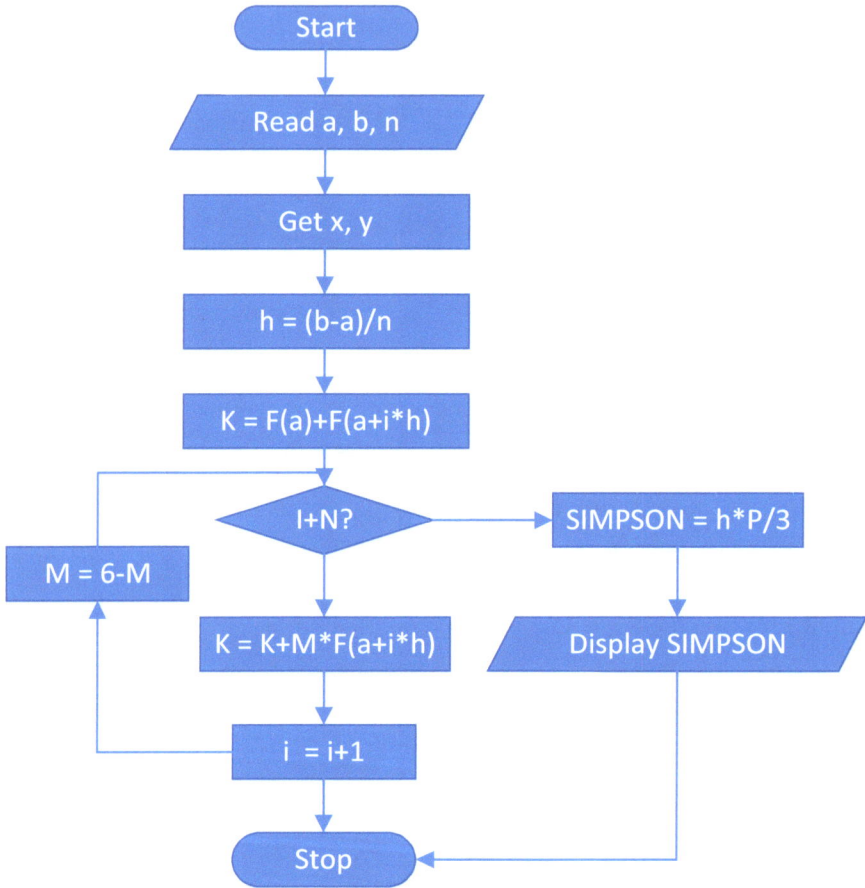

Fig. (5.13). Simpson's 1/3 method Flow chart.

Examples regarding Simson's 1/3 method can be illustrated from the same problem faced in the Trapezoidal method, which is given below:

Example 5.3.3: Using Simpson's 1/3 method, determine the voltage across the capacitor by assuming $v(0) = 0$ V from $t = 0$ to $t = 1$ second for the following circuit.

   (a) Time step 0.5s   (b) Time step 0.1s

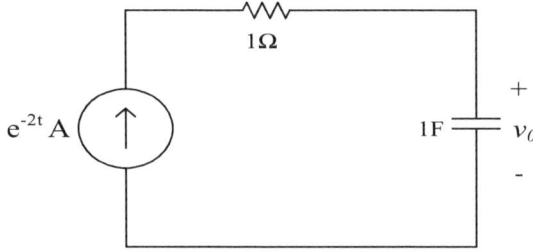

Solution:

At first, we need to know the voltage across the capacitor,

The equation of the capacitor voltage is,

$$v_0 = \frac{1}{C} \int_0^1 i \, dt$$

$$\text{Or, } v_0 = \int_0^1 e^{-2t} \, dt$$

The analytical solution of this integral is $v_0 = 0.432332$ V

Now for step size h=0.5

The table for the given problem is

t	v0
0	1
0.5	0.36788
1	0.13534

Now, the Simpson's 1/3 method gives

$$\int_{x_0}^{x_n} y \, dx = \frac{h}{3}(y_0 + 4(y_1 + y_3 + y_5 + \ldots + y_{n-1}) + 2(y_2 + y_4 + y_6 + \ldots + y_{n-2}) + y_n)$$

So, $v_0 = \int_0^1 e^{-2t} dt = \dfrac{0.5}{3}(1 + 4*0.36788 + 0.13534)$

$v_0 = 0.43447$ V

Now for step size h=0.1

The data table becomes

t	v_0
0	1
0.1	0.81873
0.2	0.67032
0.3	0.54881
0.4	0.44933
0.5	0.36788
0.6	0.30119
0.7	0.2466
0.8	0.2019
0.9	0.1653
1	0.13534

Now, the Simpson's 1/3 method gives

$$\int_{x_0}^{x_n} y\,dx = \frac{h}{3}(y_0 + 4(y_1 + y_3 + y_5 + \ldots + y_{n-1}) + 2(y_2 + y_4 + y_6 + \ldots + y_{n-2}) + y_n)$$

So, $v_0 = \int_0^1 e^{-2t} dt = \dfrac{0.1}{3}(1 + 4*(0.81873 + .54881 + .36788 + .2466 + .1653)$

$$+2*(.67032 + .44933 + .30119 + .2019) + 0.13534)$$

$$v_0 = 0.432336 \text{ volts}$$

It is seen from the above results that Simpson's 1/3 method is a much better approach than the trapezoidal method as it fitted the data with a higher degree of polynomial, as shown in Figs. (**5.11**) and (**5.12**).

It is also noticeable that the error is reduced greatly by making the step size smaller as it integrates the function by taking smaller parts each time. So, depending on the accuracy desired, one should choose the step size as required as it can increase the calculations greatly.

Example 5.3.4: The outflow concentration from a reactor is measured at a number of time instants over 24 hrs period.

Time (h)	0	4	8	12	16	20	24
C (mg/L)	1	2	2.3	4	5.5	3	1.2

The flowrate of the outflow in m^3 / s can be computed with the following equation:

$$Q(t) = 20t \times 10\sin(\frac{2\pi}{24}\{t-10\})$$

Use the Simpson's 1/3 Method to determine the flow weighted average concentration living the reactor over the 24 hour period,

$$\text{Using, } \overline{C} = \frac{\int_0^t Q(t)C(t)\,dt}{\int_0^t Q(t)\,dt}$$

Solution:

For the integration, the required table becomes

T(hr.)	C(mg/L)	Q(t) (m³/s)	Q(t)C(t)
0	1	19.54	19.54
4	2	19.72	39.44
8	2.3	19.90	45.77

12	4	20.09	80.36
16	5.5	20.27	111.485
20	3	20.45	61.35
24	1.2	20.63	24.756

Now the Simpson's 1/3 method gives

$$\int_{x_0}^{x_n} y\,dx = \frac{h}{3}(y_0 + 4(y_1 + y_3 + y_5 + \ldots + y_{n-1}) + 2(y_2 + y_4 + y_6 + \ldots + y_{n-2}) + y_n)$$

$$\int_0^{24} Q(t)C(t)\,dt = \frac{4}{3}[19.54 + 4*(39.44 + 80.36 + 61.35) + 2*(45.77 + 111.485) + 24.756]$$

$$= 1444.541333$$

$$\int_0^{24} Q(t)\,dt = \frac{4}{3}[19.54 + 4*(19.72 + 20.09 + 20.45) + 2*(19.90 + 20.27) + 20.63]$$

$$= 482.067$$

$$\therefore \overline{C} = \frac{1444.541333}{482.067} = 2.996557186. \quad {}_{(Ans)}$$

Simpson's 3/8 Method

In Equ. **5.3.7,** if we put $n = 3$, then we can obtain the general formula for Simpson's $\frac{3}{8}$ method. We obtain the equation as follows:

$$I = 3h\left[y_0 + \frac{3}{2}\Delta y_0 + \frac{3}{4}\Delta^2 y_0 + \frac{1}{8}\Delta^3 y_0\right] \qquad (5.3.22)$$

As the name suggests, we take 3/8 in common and get

$$I = \frac{3}{8} h[8y_0 + 12\Delta y_0 + 6\Delta^2 y_0 + \Delta^3 y_0]$$

Putting the values of $\Delta y_0, \Delta^2 y_0, \Delta^3 y_0$ and simplifying, we get

$$\int_{x_0}^{x_3} y dx = \frac{3}{8} h[y_0 + 3y_1 + 3y_2 + y_3] \qquad (5.3.23)$$

Similarly,

$$\int_{x_3}^{x_6} y dx = \frac{3}{8} h[y_3 + 3y_4 + 3y_5 + y_6] \qquad (5.3.24)$$

Preceding in a similar way, we obtain the formula for the remaining intervals, i.e., for $[x_6, x_9], [x_9, x_{12}] \ldots \ldots \ldots [x_{n-3}, x_n]$ and by adding them up, we get

$$\int_{x_0}^{x_n} y dx = \frac{3}{8} h[y_0 + 3(y_1 + y_2 + y_4 + y_5 + \ldots + y_{n-2} + y_{n-1}) + 2(y_3 + y_6 + \ldots + y_{n-3}) + y_n] \qquad (5.3.25)$$

Eq. **5.3.25** is called the "Simpson's 3/8 formula" for solving any integral in a given interval.

Algorithm:

1. Given a function f(x):

2. (Get user inputs)

3. Input a,b=endpoints of interval n=number of intervals(Even) (Do the integration)

4. Set sum=0.

5. Begin For i= 1 to n -1

6. Set x =a + h*i.

7. If i%3=0 Then Set sum=sum+2*f(x)

8. Else Set sum=sum+3*f(x) End

9. Set sum = sum + f(a)+f(b)

10. Set ans = sum*(3h/8)

11. End

Flowchart for Simpson's 3/8 Method is shown in Fig. **(5.14).**

Flowchart:

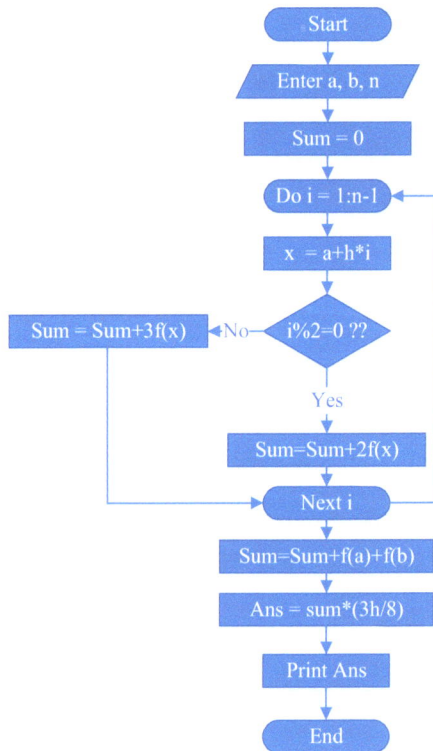

Fig. (5.14). Flowchart for Simpson's 3/8 Method.

The same example can be solved by using this method to understand this method better.

Example 5.3.5: Solve previous Example 5.3.3 using "Simpson's 3/8 method" with time step 0.1s.

Solution:

The functional value table for time step 0.1 is given as before

t	v_0
0	1
0.1	0.81873
0.2	0.67032
0.3	0.54881
0.4	0.44933
0.5	0.36788
0.6	0.30119
0.7	0.2466
0.8	0.2019
0.9	0.1653
1	0.13534

Using equation **5.3.25**, we get,

$$v_0 = \int_0^1 i\,dt = \frac{3}{8}h[y_0 + 3(y_1 + y_2 + y_4 + y_5 + y_7 + y_8) + 2(y_3 + y_6 + y_9) + y_{10}]$$

$$\therefore v_0 = 0.42863325 \text{ volts}$$

Example 5.3.6: Solve the Example 5.3.4 using Simpson's 3/8 method.

Solution:

By assuming the same time step, we get the table

T(hr)	C(mg/L)	Q(t) (m³/s)	Q(t)C(t)
0	1	19.54	19.54

4	2	19.72	39.44
8	2.3	19.90	45.77
12	4	20.09	80.36
16	5.5	20.27	111.485
20	3	20.45	61.35
24	1.2	20.63	24.756

Now, By Simpson's 3/8 Method, we have

$$\int_0^{24} Q(t)C(t)dt = \frac{3}{8} h \left[y_0 + 3\left(y_1 + y_2 + y_4 + y_5\right) + 2y_3 + y_6 \right]$$

$$= 1468.7265$$

$$\int_0^{24} Q(t)dt = \frac{3}{8} h \left[y_0 + 3\left(y_1 + y_2 + y_4 + y_5\right) + 2y_3 + y_6 \right]$$

$$= 482.055$$

$$\text{So, C} = \frac{1468.7265}{482.055} = 3.0468 \text{ mg/L}$$

Error in Simpson's 3/8 Method

Using the same approach as the trapezoidal method, it can be shown that the error in Simpson's 3/8 method is

$$\text{E} = -\frac{3}{80} h^5 (y_0^{iv} + y_1^{iv} + y_2^{iv} + \dots\dots) \qquad \textbf{(5.3.26)}$$

This method is not so accurate as of the trapezoidal method. Note that the error for the time step 0.1 in the first example was 0.333% but using Simpson's method, we can determine the error as follows:

$$\text{E} = \frac{0.42863325 - 0.43233235\,84}{0.43233235\,84} \times 100\% = 0.856\%$$

We can see that the same solution using Simpson's method gives an apparently large error which is not expected. However, this method can be useful in some specific fields of calculation. The basic conclusion is that the smaller the value of h, the more accurate the result will be. Both error equations imply this fact.

Double Integration

Double integral I of a function $f(x,y)$ of two independent variables x and y over a region R in xy plane is defined as the limit

$$I = \lim_{m,n\to\infty} \sum_{j=1}^{m}\sum_{i=1}^{n} f\left(x_i, y_j\right)\Delta x \Delta y \tag{5.3.27}$$

when the limit exists, m and n are the numbers of segments of the interval $[a,b]$ and $[c,d]$, respectively. It can also be written as

$$I = \iint_R f(x, y)dxdy \tag{5.3.28}$$

Geometrically, it represents the volume between $f(x,y)$ and the region R, where $R = [a,b]\times[c,d]$, which is illustrated in Fig. (**5.15**).

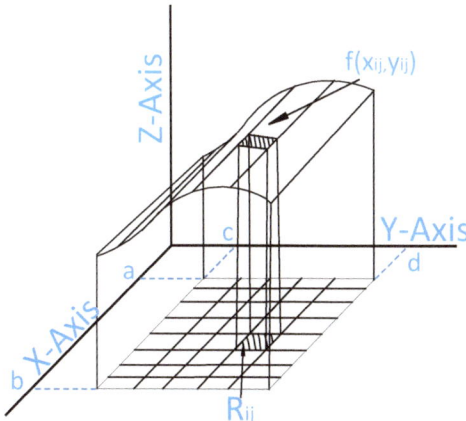

Fig. (5.15). Double integration geometrically.

Eq. **(5.3.28)** is useful for the evaluation of indefinite double integral, while equation **5.3.27** is useful for definite integral. Double integrals are computed as iterated integrals. Thus, the integral in one of the dimensions is evaluated first. The result of this first integration is integrated into the second dimension. Mathematically, this is expressed as:

$$I = \int_c^d \left(\int_a^b f(x,y)dx \right) dy = \int_a^b \left(\int_c^d f(x,y)dy \right) dx \qquad (5.3.29)$$

Numerical evaluation of definite integral from a tabulated value of $f(x,y)$ is only considered further.

When $m \to \infty$ and $n \to \infty$ in Eq. 5.3.29, the exact solution is found. But in all practical cases, m and n are finite, so the double integral is approximated using multiple-segment trapezoidal or Simpson's rule.

A numerical double integral would be based on the same idea as indefinite integrals. First, methods like the multiple-segment trapezoidal or Simpson's rule would be applied in the first dimension, with each value of the second dimension held constant. Then, the method would be applied to integrate the second dimension.

It is visually illustrated in Fig. **(5.16)** that volumes of cuboids are used to approximate the volume under $f(x,y)$ surface in region R.

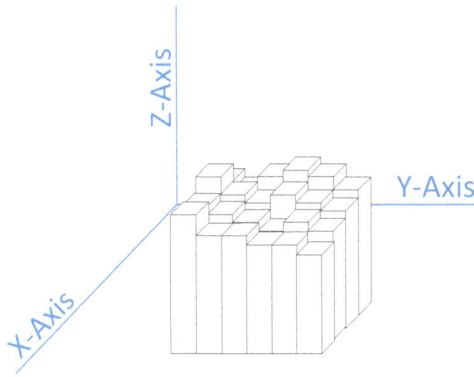

Fig. (**5.16**). Numerical approximation of double integral using cuboids.

Trapezoidal Rule

The trapezoidal rule can be simply applied repeatedly to evaluate double integrals.

The below integral can be evaluated using trapezoidal rule twice

$$I = \int_{y_1}^{y_m} \left(\int_{x_1}^{x_n} f\left(x_i, y_j\right) dx \right) dy \qquad (5.3.30)$$

where,

$x_i = x + ih$

$y_j = y + jk$

$i = 0,1,2,3 \ldots\ldots\ldots\ldots\ldots$

$j = 0,1,2,3 \ldots\ldots\ldots\ldots\ldots$

$h = \dfrac{b-a}{n}$

$k = \dfrac{d-c}{m}$

A shorter notation can be used writing $f\left(x_i, y_j\right) = f_{i,j}$

Using trapezoidal rule once

$$I = \int_{y_1}^{y_m} \left(\frac{h}{2} \left[f_{1,j} + 2\left(f_{2,j} + f_{3,j} + \ldots + f_{n-1,j} \right) + f_{n,j} \right] \right) dy$$

Again using the trapezoidal rule on every term

$$I = \frac{hk}{4} \left(2 \begin{array}{l} \left[f_{1,1} + 2\left(f_{1,2} + f_{1,3} + \ldots + f_{1,m-1} \right) + f_{1,m} \right] + \\ \left(\begin{array}{l} \left[f_{2,1} + 2\left(f_{2,2} + f_{2,3} + \ldots + f_{2,m-1} \right) + f_{2,m} \right] + \ldots \\ + \left[f_{n-1,1} + 2\left(f_{n-1,2} + f_{n-1,3} + \ldots + f_{n-1,m-1} \right) + f_{n-1,m} \right] \end{array} \right) \\ + \left[f_{n,1} + 2\left(f_{n,2} + f_{n,3} + \ldots + f_{n,m-1} \right) + f_{n,m} \right] \end{array} \right)$$

Or

$$I = \frac{hk}{4} \left(\begin{array}{l} f_{1,1} + f_{1,m} + f_{n,1} + f_{n,m} \\ +2 \left(\begin{array}{l} \left[f_{1,2} + f_{1,3} + \ldots + f_{1,m-1} \right] + \left[f_{n,2} + f_{n,3} + \ldots + f_{n,m-1} \right] \\ + \left[f_{2,1} + f_{3,1} + \ldots + f_{n-1,1} \right] + \left[f_{2,m} + f_{3,m} + \ldots + f_{n-1,m} \right] \end{array} \right) \\ +4 \sum_{\substack{p=2 \\ q=2}}^{\substack{p=m-1 \\ q=n-1}} f_{p,q} \end{array} \right) \tag{5.3.31}$$

This equation can be easily expressed in terms of the matrix. The coefficients for a single variable trapezoidal rule are

$$C1 = [1 \quad 2 \quad . \quad . \quad . \quad 2 \quad 1]$$

Then, the coefficients matrix for two-variable trapezoidal rule is

$$C = \begin{bmatrix} 1 & 2 & 2 & . & 2 & 1 \\ 2 & 4 & 4 & 4 & 4 & 2 \\ . & 4 & 4 & 4 & 4 & 2 \\ 2 & 4 & 4 & 4 & 4 & . \\ 2 & 4 & 4 & 4 & 4 & 2 \\ 1 & 2 & . & 2 & 2 & 1 \end{bmatrix}$$

Where

$$c_{ij} = cl_i \times cl^T{}_j \tag{5.3.32}$$

i = row number

j = column number

If the functional values are given in matrix D, then

$$I = \frac{hk}{4} \sum_{j=1}^{m} \sum_{i=1}^{n} c_{ij} \times d_{ij} \tag{5.3.33}$$

Simpson's 1/3 Rule

Consider the integral

$$I = \int_{y_{j-1}}^{y_{j+1}} \left(\int_{x_{i-1}}^{x_{i+1}} f\left(x_i, y_j\right) dx \right) dy \tag{5.3.34}$$

It can be evaluated using Simpson's 1/3 rule twice like previously done using the trapezoidal rule.

$$I = \int_{y_{j-1}}^{y_{j+1}} \left(\frac{h}{3} \left[f\left(x_{i-1}, y\right) + 4f\left(x_i, y\right) + f\left(x_{i+1}, y\right) \right] \right) dy$$

$$I = \frac{hk}{9} \begin{pmatrix} \left[f\left(x_{i-1}, y_{j-1}\right) + 4f\left(x_{i-1}, y_j\right) + f\left(x_{i-1}, y_{j+1}\right) \right] \\ + \left[f\left(x_i, y_{j-1}\right) + 4f\left(x_i, y_j\right) + f\left(x_i, y_{j+1}\right) \right] \\ + \left[f\left(x_{i+1}, y_{j-1}\right) + 4f\left(x_{i+1}, y_j\right) + f\left(x_{i+1}, y_{j+1}\right) \right] \end{pmatrix}$$

$$I = \frac{hk}{9} \begin{pmatrix} f\left(x_{i-1}, y_{j-1}\right) + f\left(x_{i-1}, y_{j+1}\right) + f\left(x_{i+1}, y_{j-1}\right) + f\left(x_{i+1}, y_{j+1}\right) \\ + 4\left[f\left(x_i, y_{j-1}\right) + 4f\left(x_i, y_{j+1}\right) + f\left(x_{i-1}, y_j\right) + f\left(x_{i+1}, y_j\right) \right] + 16f\left(x_i, y_j\right) \end{pmatrix} \tag{5.3.35}$$

This equation can also be expressed in terms of the matrix in a generalized form. The coefficients for single variable Simpson's 1/3 rule are

$$C1 = [1 \quad 4 \quad 2 \quad 4 \quad \ldots\ldots \quad 2 \quad 1]$$

Then, the coefficients matrix for two variable Simpson's 1/3 rule is

Where

$$c_{ij} = c1_i \times c1^T_j \tag{5.3.36}$$

i=row number

j=column number

If the function values are given in matrix D, then

$$I = \frac{hk}{9} \sum_{j=1}^{m} \sum_{i=1}^{n} c_{ij} \times d_{ij} \tag{5.3.37}$$

For example, if m=4 and n=6, then the coefficient matrix would be

$$C = \begin{bmatrix} 1 & 4 & 2 & 4 & 2 & 1 \\ 4 & 16 & 8 & 16 & 8 & 4 \\ 2 & 8 & 4 & 8 & 4 & 2 \\ 1 & 4 & 2 & 4 & 2 & 1 \end{bmatrix}$$

The matrix forms are used to quickly calculate the integral.

A generalized algorithm using the matrix methods is described below and the flowchart of the double integration is shown in Fig. (**5.17**).

Algorithm:

1. Make an m by n data matrix D from the functional value at different points, where different columns represent different values of x and different rows represent different values of y.
2. Make an m by n coefficient matrix C using equation **5.3.32** or **5.3.36** for the different methods. Here, the maximum value of I is m and the maximum value of j is n.
3. Perform element-wise multiplication of C and D, then add all the elements.
4. Determine the integral form equation **5.3.33** or **5.3.37**.

Flowchart:

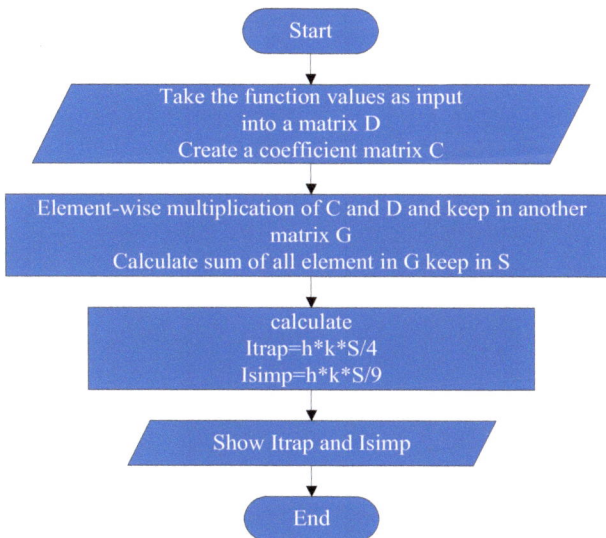

Fig. (**5.17**). Flowchart for double integration.

Example 5.3.7: Calculate the electric charge due to a cuboid object having a face with irregular surface $f(x,y)$ (surface without exact mathematical equation or with complicated equation) on one side and planes on all other sides. The charge density in the object is 1 C/m³. The value of $f(x,y)$ at different grid points is given in the table. [$f(x,y)=xy$]

$y \downarrow x \rightarrow$	1	1.2	1.4	1.6	1.8	2
1	1	1.2	1.4	1.6	1.8	2
1.25	1.25	1.5	1.75	2	2.25	2.5
1.5	1.5	1.8	2.1	2.4	2.7	3
1.75	1.75	2.1	2.45	2.8	3.15	3.5
2	2	2.4	2.8	3.2	3.6	4

Solution:

Here,

$$\rho_v = 1 \ \text{C/m}^3$$

m=5

n=6

h=0.2

k=0.25

We know

$$Q = V \rho_v$$

$$V = \iint f(x,y)dxdy = k1 \sum_{j=1}^{m} \sum_{i=1}^{n} c_{ij} \times d_{ij}$$

Trapezoidal Rule

For trapezoidal rule,

k1=hk/4=0.2*0.25/4=0.0125

$$D= \begin{bmatrix} 1.0000 & 1.2000 & 1.4000 & 1.6000 & 1.8000 & 2.0000 \\ 1.2500 & 1.5000 & 1.7500 & 2.0000 & 2.2500 & 2.5000 \\ 1.5000 & 1.8000 & 2.1000 & 2.4000 & 2.7000 & 3.0000 \\ 1.7500 & 2.1000 & 2.4500 & 2.8000 & 3.1500 & 3.5000 \\ 2.0000 & 2.4000 & 2.8000 & 3.2000 & 3.6000 & 4.0000 \end{bmatrix}$$

Using Eq. **5.3.32**

$$C= \begin{bmatrix} 1 & 2 & 2 & 2 & 2 & 1 \\ 2 & 4 & 4 & 4 & 4 & 2 \\ 2 & 4 & 4 & 4 & 4 & 2 \\ 2 & 4 & 4 & 4 & 4 & 2 \\ 1 & 2 & 2 & 2 & 2 & 1 \end{bmatrix}$$

Then C.*D $= \begin{bmatrix} 1.0000 & 2.4000 & 2.8000 & 3.2000 & 3.6000 & 2.0000 \\ 2.5000 & 6.0000 & 7.0000 & 8.0000 & 9.0000 & 5.0000 \\ 3.0000 & 7.2000 & 8.4000 & 9.6000 & 10.8000 & 6.0000 \\ 3.5000 & 8.4000 & 9.8000 & 11.200 & 12.6000 & 7.0000 \\ 2.0000 & 4.8000 & 5.6000 & 6.4000 & 7.2000 & 4.0000 \end{bmatrix}$

So,

$$\sum_{j=1}^{m}\sum_{i=1}^{n} c_{ij} \times d_{ij} = 180 \text{ [adding all the elements of C.*D]}$$

Then,

$$V = 0.0125 \times 180 = 2.25 \text{ m}^3$$

$$Q = 2.25 \times 1 = 2.25\,C \ \ (\text{Ans})$$

Simpson's 1/3 Rule:

k1=hk/9=0.2*0.25/9=0.0056

D is the same as before but

Using Eq. **5.3.36**

$$C = \begin{bmatrix} 1 & 4 & 2 & 4 & 2 & 1 \\ 4 & 16 & 8 & 16 & 8 & 4 \\ 2 & 8 & 4 & 8 & 4 & 2 \\ 4 & 16 & 8 & 16 & 8 & 4 \\ 1 & 4 & 2 & 4 & 2 & 1 \end{bmatrix}$$

Then $C.*D = \begin{bmatrix} 1.0000 & 4.8000 & 2.8000 & 6.4000 & 3.6000 & 2.0000 \\ 5.0000 & 24.0000 & 14.0000 & 32.0000 & 18.0000 & 10.0000 \\ 3.0000 & 14.4000 & 8.4000 & 19.2000 & 10.8000 & 6.0000 \\ 7.0000 & 33.6000 & 19.6000 & 44.8000 & 25.2000 & 14.0000 \\ 2.0000 & 9.6000 & 5.6000 & 12.8000 & 7.2000 & 4.0000 \end{bmatrix}$

So,

$$\sum_{j=1}^{m}\sum_{i=1}^{n} c_{ij} \times d_{ij} = 370.8 \text{ m}^3 \ [\text{adding all the elements of C.*D}]$$

Then,

$$V = 0.0056 \times 370.8 = 2.0765 \text{ m}^3$$

$$Q = 2.0765 \times 1 = 2.0765 \text{ C (Ans)}$$

Example 5.3.8: Calculate the total magnetic flux **f** from the values of magnetic flux density B flowing through the region $R = (0,1) \times (0,1)$, given in the below table.

$$[\, B = \mathrm{Re}\left(\sqrt{1 - x^2 - y^2}\right)\,]$$

$y \downarrow x \rightarrow$	0	0.2	0.4	0.6	0.8	1
0	1	0.979796	0.916515	0.8	0.6	0
0.2	0.979796	0.959166	0.894427	0.774597	0.565685	0
0.4	0.916515	0.894427	0.824621	0.69282	0.447214	0
0.6	0.8	0.774597	0.69282	0.52915	0	0

| 0.8 | 0.6 | 0.565685 | 0.447214 | 0 | 0 | 0 |
| 1 | 0 | 0 | 0 | 0 | 0 | 0 |

Solution:

Here,

m=5

n=6

h=0.2

k=0.2

We knows

$$\varphi = \iint B dx dy$$

$$\varphi = \iint B dx dy = k1 \sum_{j=1}^{m} \sum_{i=1}^{n} c_{ij} \times d_{ij}$$

Trapezoidal Rule:

For trapezoidal rule,

k1=hk/4=0.2*0.2/4=0.01

$$D = \begin{bmatrix} 1.0000 & 0.9798 & 0.9165 & 0.8000 & 0.6000 & 0 \\ 0.9798 & 0.9592 & 0.8944 & 0.7746 & 0.5657 & 0 \\ 0.9165 & 0.8944 & 0.8246 & 0.6928 & 0.4472 & 0 \\ 0.8000 & 0.7746 & 0.6928 & 0.5292 & 0 & 0 \\ 0.6000 & 0.5657 & 0.4472 & 0 & 0 & 0 \\ 0 & 0 & 0 & 0 & 0 & 0 \end{bmatrix}$$

Using Eq. **5.3.32**

$$C = \begin{bmatrix} 1 & 2 & 2 & 2 & 2 & 1 \\ 2 & 4 & 4 & 4 & 4 & 2 \\ 2 & 4 & 4 & 4 & 4 & 2 \\ 2 & 4 & 4 & 4 & 4 & 2 \\ 2 & 4 & 4 & 4 & 4 & 2 \\ 1 & 2 & 2 & 2 & 2 & 1 \end{bmatrix}$$

Then $C.*D =$
$$\begin{bmatrix} 1.0000 & 1.9596 & 1.8330 & 1.6000 & 1.2000 & 0 \\ 1.9596 & 3.8367 & 3.5777 & 3.0984 & 2.2627 & 0 \\ 1.8330 & 3.5777 & 3.2985 & 2.7713 & 1.7889 & 0 \\ 1.6000 & 3.0984 & 2.7713 & 2.1166 & 0 & 0 \\ 1.2000 & 2.2627 & 1.7889 & 0 & 0 & 0 \\ 0 & 0 & 0 & 0 & 0 & 0 \end{bmatrix}$$

So,

$$\sum_{j=1}^{m}\sum_{i=1}^{n} c_{ij} \times d_{ij} = 50.4349 \quad \text{[adding all the elements of C.*D]}$$

Then,

$$\varphi = 0.01 \times 50.4349 = 0.5044 \text{ Wb (Ans)}$$

Simpson's 1/3 Rule:

$k1 = hk/9 = 0.2*0.2/9 = 0.0044$

D is the same as before but

Using Eq. **5.3.36**

$$C = \begin{bmatrix} 1 & 4 & 2 & 4 & 2 & 1 \\ 4 & 16 & 8 & 16 & 8 & 4 \\ 2 & 8 & 4 & 8 & 4 & 2 \\ 4 & 16 & 8 & 16 & 8 & 4 \\ 2 & 8 & 4 & 8 & 4 & 2 \\ 1 & 4 & 2 & 4 & 2 & 1 \end{bmatrix}$$

$$\text{Then } C.*D = \begin{bmatrix} 1.0000 & 3.9192 & 1.8330 & 3.2000 & 1.2000 & 0 \\ 3.9192 & 15.3467 & 7.1554 & 12.3935 & 4.5255 & 0 \\ 1.8330 & 7.1554 & 3.2985 & 5.5426 & 1.7889 & 0 \\ 3.2000 & 12.3935 & 5.5426 & 8.4664 & 0 & 0 \\ 1.2000 & 4.5255 & 1.7889 & 0 & 0 & 0 \\ 0 & 0 & 0 & 0 & 0 & 0 \end{bmatrix}$$

So,

$$\sum_{j=1}^{m}\sum_{i=1}^{n} c_{ij} \times d_{ij} = 111.2277 \text{ [adding all the elements of C.*D]}$$

Then,

$$\varphi = 0.0044 \times 111.2277 = 0.4894 \text{ Wb (Ans)}$$

Romberg Integration

The Romberg Integration method uses the Richardson extrapolation formula to improve the approximations obtained by the finite-difference methods, *i.e.*, methods that use Trapezoidal rule, Simpson's 1/3-Rule or Simpson's 3/8 Rule, *etc.*

$$I_o = \int_{x_i}^{x_f} y \, dx \tag{5.3.38}$$

Assuming that I_h the result obtained by taking h as subinterval, E_h the value of error that occurred for this subinterval and I_o the result obtained by taking the subinterval infinitesimally small, we can write

$$I_h = I_o + E_h$$

Thus, if we use two different values of increment h and calculate it, we get

$$\text{for subinterval of } h_1, \ I_o = I_{h_1} - E_{h_1}$$

and

$$\text{for subinterval of } h_2, \ I_o = I_{h_2} - E_{h_2} \tag{5.3.39}$$

This implies,

$$I_{h_1} - E_{h_1} = I_{h_2} - E_{h_2}$$

$$E_{h_2} - E_{h_1} = I_{h_2} - I_{h_1} \tag{5.3.40}$$

Assuming that I_{h_1} & I_{h_2} were evaluated in trapezoidal method, then the errors can be written as

$$E_{h_1} = -\frac{x_n - x_0}{12} h_1{}^2 y''(\bar{x})$$

and

$$E_{h_2} = -\frac{x_n - x_0}{12} h_2{}^2 y''(\bar{\bar{x}})$$

Both the terms $y''(\bar{x})$ & $y''(\bar{\bar{x}})$ are taken as the largest value of $y''(x)$, so these two terms can be assumed to be almost equal. We, therefore, have

$$\frac{E_{h_1}}{E_{h_2}} \cong \frac{h_1{}^2}{h_2{}^2}$$

Hence,

$$\frac{E_{h_2}}{E_{h_2} - E_{h_1}} \cong \frac{h_2{}^2}{h_2{}^2 - h_1{}^2} \tag{5.3.41}$$

Using Eqs. **5.3.40** & **5.3.41**

$$E_{h_2} \cong \frac{h_2{}^2}{h_2{}^2 - h_1{}^2}\left(I_{h_2} - I_{h_1}\right)$$

By substituting the value E_{h_2} in Eq. **5.3.39**,

$$I_o = I_{h_2} - E_{h_2}$$

$$= I_{h_2} - \frac{h_2{}^2}{h_2{}^2 - h_1{}^2}\left(I_{h_2} - I_{h_1}\right)$$

$$= \frac{I_{h_1}h_2{}^2 - I_{h_2}h_1{}^2}{h_2{}^2 - h_1{}^2}$$

If the errors are of the same sign and decrease monotonically, this I_o would be closer to the actual value of the integration.

If we now set

$$h_1 = h, \qquad h_{2=}\frac{h}{2}$$

Therefore,

$$I_{h_1} = I(h), \qquad I_{h_2} = I\left(\frac{h}{2}\right)$$

The equation can be rewritten as

$$I\left(h,\frac{h}{2}\right) \cong \frac{1}{3}\left[4I\left(\frac{h}{2}\right) - I(h)\right] \tag{5.3.42}$$

To get a clearer idea of how these assumptions are made is described in Fig. (**5.18**),

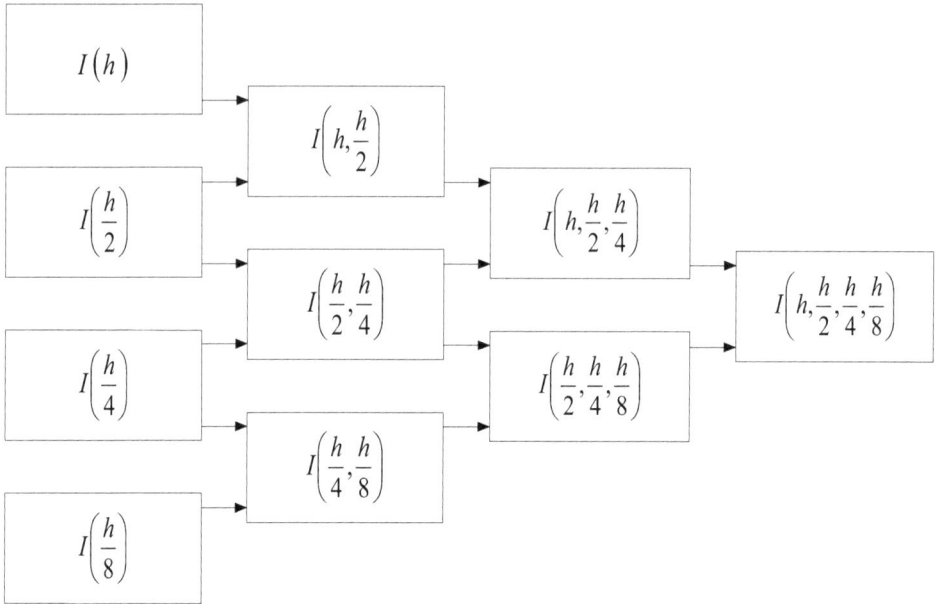

Fig. (5.18). Process of Romberg Integration.

Clearly, the number of calculations depends on the requirement of the accuracy of the result.

Note that the new estimate found in equation 5 is of $O(h^4)$. If two new estimates are to be used to get a more appropriate result, then it will be of $O(h^6)$, $O(h^8)$ and so on.

For example,

$$I\left(h,\frac{h}{2},\frac{h}{4}\right) \cong \frac{1}{15}\left[16I\left(h,\frac{h}{2}\right) - I\left(\frac{h}{2},\frac{h}{4}\right)\right]$$

Again,

$$I\left(h,\frac{h}{2},\frac{h}{4},\frac{h}{8}\right) \cong \frac{1}{63}\left[64I\left(h,\frac{h}{2},\frac{h}{4}\right) - I\left(\frac{h}{2},\frac{h}{4},\frac{h}{8}\right)\right]$$

Flowchart for the Romberg integration is Fig. (**5.19**).

Flowchart:

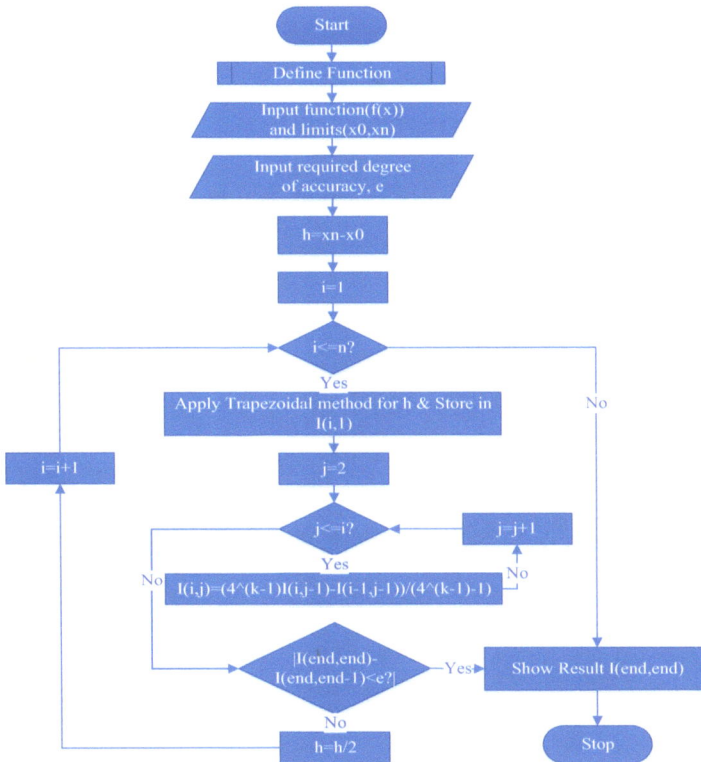

Fig. (5.19). Flowchart for the Romberg integration.

Algorithm:

1. Start
2. Define Function
3. Input function, upper limit, lower limit & maximum error
4. Take h initially as the difference between two limits
5. Take a value i=1
6. Apply Trapezoidal method and store in I (i,1)
7. Take value j=2:i
8. Apply formula I(i,j)=(4^(k-1)I(i,j-1)-I(i-1,j-1))/(4^(k-1)-1) for all the values of j
9. Check if the error is lower than required. If yes, go to step 12, else go to step 10
10. Divide h by 2 and store in h
11. Increment value of i & go to step 6
12. Show I(end,end) as a result
13. Stop

CONSENT FOR PUBLICATION

Not applicable.

CONFLICT OF INTEREST

The author declares no conflict of interest, financial or otherwise.

ACKNOWLEDGEMENT

Declared none.

REFERENCES

[1] D. D. Moursund, C. S. Duris, *Elementary theory and applications of numerical analysis.* McGraw-Hill, New York, 1967.
[2] S. C. Chapra, R. P. Canale, *Numerical methods for engineers.* McGraw-Hill Education, New York, USA.
[3] Y. Jaluria, *Computer Methods for Engineering.* Allyn and Bacon, 1988.
[4] V. Zalizniak, *Essentials of Scientific Computing : Numerical Methods for Science and Engineering.* Chichester, UK: Horwood Pub., 2008.
[5] S. S. Sastry, Introductory Methods of Numerical Analysis, PHI Learning, Delhi, 2013.
[6] E Balagurusamy, Numerical Methods, Tata McGraw Hill Education, New Delhi, 2014.

EXERCISES

1. The distance travelled by an electron inside a conductor at different times (t sec) are given below:

T (sec)	0.0	0.2	0.4	0.6	0.8	1
	3.01	**3.17**	**3.28**	**3.36**	**3.42**	**3.39**

Find the velocity of the electron at time t=0.4 sec.

2. The switch in the following circuit is closed at t=0. After closing the switch, recorder currents for different times are tabulated as follows:

t (sec)	0	2	4	6	8

i (imp)	0	0.632	0.864	0.992	0.174

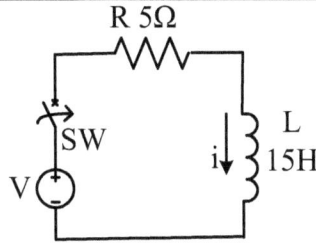

Obtain the source voltage at time t = 2 sec using the numerical method.

3. The input of a differentiator circuit is given by $g(x) = x^2 + 3x$. Calculate the output at $x = 2$ for $h = 0.1$ by using the forward and central difference quotient and compare the results.

4. The distances travelled by an electron inside a conductor a different times (sec) are given below:

t	0.00	0.10	0.20	0.30	0.40	0.50	0.60
x	2.01	2.19	2.29	2.39	2.45	3.38	3.31

Find out the acceleration of the electron at t=0.2

5. Using all the methods applicable, determine the voltage across the capacitor by assuming $v(0)$ = 0 Volts from $t = 0$ to $t = 1$ second for the following circuit.

    (a) Time step 0.2 seconds   (b) Time step 0.01 seconds

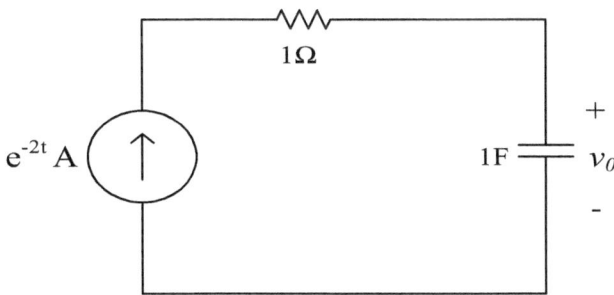

CHAPTER 6

Numerical Solution of Ordinary Differential Equation

Md. Shamim Anower[1]*, Md. Rashidul Islam[1] and Md. Mahabubur Rahman[1]

[1] Department of Electrical & Computer Engineering, Rajshahi University of Engineering & Technology, Bangladesh

Abstract: Practically, the engineer's deal with a lot of problems that can be expressed mathematically by ordinary differential equations (ODEs). These ODEs can be solved using both direct and iterative methods. The latter is popular as, in this case, the solution techniques are based only on the basic arithmetic operations. In this chapter, at first, we studied ordinary differential equations. Secondly, fundamental theories for the solutions of these differential equations are discussed. Then, various numerical solution techniques are explained in this regard. At the end of each technique, solutions to various engineering problems are discussed.

Keywords: Ordinary differential equation, Taylor series method, Picard's method, Euler's method, Runge-Kutta Finite difference formulae, Predictor-Corrector method, solving simultaneous first order ODEs, solving higher order ODEs.

INTRODUCTION

Physical systems, mathematically modeled by differential equations, are very common in different branches of engineering. Two types of differential equations, ordinary and partial, are obvious in practice. Ordinary differential equations contain derivatives with only one independent variable. Partial differential equations contain derivatives with more than one independent variable. In this text, we consider computational methods for solving ordinary differential equations, that is, those differential equations that have only one independent variable.

The aim of this chapter is to provide an introduction to computational methods for the solution of ordinary differential equations (ODEs). Obviously, the solution obtained in numerical method is not exact; rather, the methods discussed here give

*Corresponding author Md. Shamim Anower: Department of Electrical & Computer Engineering, Rajshahi University of Engineering & Technology, Bangladesh; E-mail: md.shamimanower@yahoo.com

Md. Masud Rana, Wei Xu, and Youguang Guo (Eds.)

approximate solutions. Ordinary differential equations frequently occur as mathematical models in many branches of science, engineering and economy. Unfortunately, it is seldom that these equations have solutions that can be expressed in closed form, so it is common to seek approximate solutions by means of numerical methods; nowadays, this can usually be achieved very inexpensively to high accuracy and with a reliable bound on the error between the analytical solution and its numerical approximation [1]. In this chapter, only minimal prerequisites are assumed. This chapter focuses on the construction of numerical algorithms for ODEs and the mathematical analysis of their behavior. This chapter begins with a brief study of initial value problems, boundary value problems and systems of such equations. The basic ideas of error analysis are also introduced here. This chapter also includes concepts of stability. For the simplicity of understanding, the same mathematical problem is solved here using every method discussed here. The final sections are devoted to various practical electrical problems which will encourage the reader to understand and apply these numerical methods in various practical situations.

ORDINARY DIFFERENTIAL EQUATIONS

Equations containing derivatives with respect to one independent variable are termed as ordinary differential equations. Ordinary differential equations are among the most important mathematical tools used in producing models in the physical sciences, biological sciences, and engineering. According to the order of the derivative, they are classified as first order and higher order ODEs. Sometimes we have to deal with simultaneous equations of first order derivatives.

Introduction

Now, consider the differential equation of the form

$$\frac{dy}{dx} = f(x, y); \text{ where, } y(x_1) = y_1. \tag{6.2.1}$$

Here, the condition in the right indicates the initial relation between dependent and independent variables. If the initial value of independent variable, $x = x_1$, the value of dependent variable, $y = y_1$.

The aforesaid equation in (6.2.1) is called first-order ordinary differential equation as it contains only the first-order derivative of the dependent variable. If in a case, the second order derivative of the dependent variable exists, it will be called second-order ordinary differential equation. Thus, the existence of higher-order derivative gives the name higher-order ordinary differential equation. A higher-order differential equation can be reformulated as a system of first-order equations. The numerical solution techniques for a first-order equation can be extended in a straightforward way to a system of first-order equations [2]. So, we will concentrate basically on the solution techniques for the first order differential equations. The solution of higher-order equations is discussed in brief at the end of the chapter.

The function in the right side of (6.2.1) may perhaps be a typical function of independent variable x or dependent variable, y or any combination of them. In practice, sometimes, we may have a table of values of this function. To solve ODEs, we require knowing at least a value of the dependent variable for a particular value of independent variable at any of the two extremities. Thus, two types of problems may arise, one is initial value problem and the other is the boundary value problem. To apply numerical methods, the solution range is also required for these problems. In the initial value problem, we will have the value of y at $x = x_1$ (initial state) and the solution is obtained in steps until the final value of the solution range ($x_1 \leq x \leq x_n$). On the other hand, if y is given at $x = x_n$ (final state) and the solution is obtained in steps until the initial value of the solution range ($x_n \geq x \geq x_1$) then, the problem is called a boundary value problem [3, 4]. We will concentrate only on the initial value problem throughout this chapter.

The Basics

Basically, there are some well-known analytical methods of solving ordinary differential equations. However, these can solve only a specific form of ODEs. But in real life, when the equations are more complex or when we will have only the tabulated values of the problem, the analytical methods are very hard and even non-applicable in some cases. In such situations, to get a solution of ODEs numerical methods are the only alternative. Numerical techniques are equally applicable for those cases where we can use analytical methods.

To solve ODEs using numerical methods, we may use any of the following two forms of solutions:

Form I: power series of the dependent variable with respect to the independent variable

Form II: tabulated values of dependent variable with respect to independent variable

Form I is practiced in Taylor's series and Picard's methods of solving ODEs. On the other hand, Form II is practiced in Euler, Runge-Kutta, Adams-Bashforth etc. methods.

Taylor Series Method

Taylor series is one of the most important formulas in mathematics. Here, satisfying certain conditions, a function can be evaluated near a certain point by the sum of infinite terms which are expressed in terms of the initial value, the power of independent variable and the function's derivatives at that point. Frequently, we use finite number of terms of this series with reasonable error in the estimation. The use of Taylor series provides the solution of ODE as follows: at first, the solution function y will be expressed in the form of Taylor series and the subsequent values of y can be obtained by direct substitution of independent variable x. Finally, we will end according to the problem statement.

Consider again the differential equation

$$y' = \frac{dy}{dx} = f(x, y)$$

$$\text{with } y = y_1 \text{ at } x = x_1$$

Numerical solution of this ODE is to obtain the values of $y(x)$ at some values of x provided where the $y(x)$ is expressed in Taylor series.

Now expanding y(x) in Taylor series around $x = x_1$ we get:

$$y(x) = y_1 + (x - x_1)y' + \frac{(x - x_1)^2}{2!}y'' + \frac{(x - x_1)^3}{3!}y''' + \cdots$$

$$= y_1 + (x - x_1)y' + \frac{(x - x_1)^2}{2}y'' + \frac{(x - x_1)^3}{6}y''' + \cdots \qquad (6.3.1)$$

If the next value of x can be expressed as $x = x_1 + h$, the value y at that x can be evaluated using the above series as follows:

$$y(x_1 + h) = y(x_1) + hy'(x_1) + \frac{h^2}{2}y''(x_1) + \frac{h^3}{6}y'''(x_1) + \cdots \qquad (6.3.2)$$

Thus, the subsequent values of the solution function y for the values of x at $x_1 + 2h$, $x_1 + 3h$, …. , $x_1 + nh$ are obtained using this Taylor series.

Although this series comprises infinite terms, we use only its finite approximations. We may get a solution using this series up to the second term where the truncation error would be of the order of h^2. An improved result can be obtained, including the third term in the above expression to extrapolate the solution function to the point $(x_1 + h)$ from x_1. Thus, further improvement in the result is obvious, including the higher order terms of the series.

Evaluating the derivatives in the solution expression is not easy all the time and enormous computational effort is required in the process. In some circumstances, when the problem statement comes with tabulated data, this scheme is not directly applicable. Prior extra computational effort is needed for numerical evaluation of the derivatives again. So it is essential to have methods which will deal with only the values of x and y and the derivatives will not be calculated. We will discuss some of such methods in the next. Here, an algorithm and a flowchart are provided to program the Taylor series method of solving ODEs.

Algorithm 6.3.1: Algorithm of Taylor Series Method to Solve ODEs

1. Start

2. Input the function

3. Enter the values of initial points x_1, y_1.

4. Enter the point x_n ,where the solution is required.

5. Enter the step size h.

6. Compute $(x_i, y_i), f'(x_i, y_i), f''(x_i, y_i)$ ....

7. Compute $y(x_i + h) = y(x_i) + hf(x_i, y_i) + \frac{h^2}{2!}f'(x_i, y_i) + \frac{h^3}{3!}f''(x_i, y_i)...$

8. Do $x_i = x_i + h$ until $x_i = x_n$ and do the step 5 again

9. Stop

Process of Taylor Series Method to Solve ODEs is shown in Fig. (6.1).

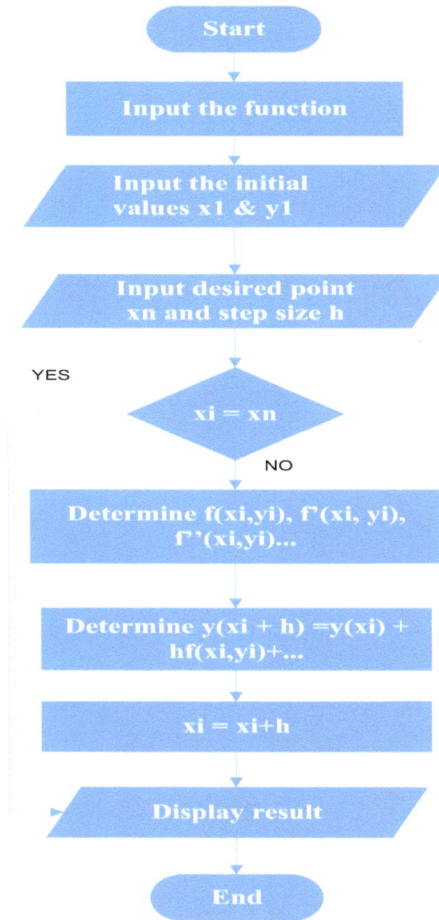

Fig. (6.1). Process of Taylor Series Method to Solve ODEs.

Following example is provided to have better understanding of the Taylor series method.

Example 6.3.1:

The output of an electric power plant for the orbiting space station uses photovoltaic cells to store energy in batteries. The charging system depends on the differential equation $\frac{dv}{dt} + 2v = 0$, with initial conditions $t_0 = 0$ & $v_0 = 1$.

Determine the value of v at $t=0.2$s using Taylor's series method correct to four decimal points.

Solution:

Here, $\frac{dv}{dt} = -2v = f(t, v)$

The initial conditions are given as $t_0 = 0$ & $v_0 = 1$.

$$v(t) = 1 + tv_0' + \frac{t^2}{2}v_0'' + \frac{t^3}{6}v_0''' + \frac{t^4}{24}v_0^{iv} + \frac{t^5}{120}v_0^v + \cdots$$

The derivatives $v_0', v_0'', v_0''', \ldots\ldots$ are obtained.

Thus,

$$v'(t) = -2v \qquad\qquad v_0' = -2$$
$$v''(t) = -2v' \qquad\qquad v_0'' = -2 * (-2) = 4$$
$$v'''(t) = -2v'' \qquad\qquad v_0''' = -2 * 4 = -8$$
$$v^{iv}(t) = -2v''' \qquad\qquad v_0^{iv} = -2 * (-8) = 16$$
$$v^v(t) = -2v^{iv} \qquad\qquad v_0^v = -2 * 16 = -32$$

Using these values, the Taylor's series becomes,

$$v(t) = 1 - 2t + \frac{t^2}{2} * 4 + \frac{t^3}{6} * (-8) + \frac{t^4}{24} * 16 + \frac{t^5}{120} * (-32) + \cdots$$

To obtain $v(0.2)$ correct to four decimal places, it is found that the terms up to t^4 should be considered and we have $v(0.2) = 0.6704$.

Picard's Method

Another method following the Form I above to find the solution of ordinary differential equation is Picard's method in which we obtain the solution using a power series.

Let us assume the given differential equation as,

$$\frac{dy}{dx} = f(x, y)$$

Integrating the above equation we get,

$$y = y_0 + \int_{x_0}^x f(x, y)dx \qquad (6.4.1)$$

The equation (6.4.1) is an integral equation which can be solved by the methods of successive approximations. The first approximation is obtained by putting y_0 for y on the right hand side of the equation (6.4.1). So, we get

$$y^{(1)} = y_0 + \int_{x_0}^x f(x, y_0)dx$$

Now using this first approximation we can obtain further approximations.

Generally,

$$y^{(n)} = y_0 + \int_{x_0}^x f(x, y^{(n-1)})dx$$

The method converges to the solution if f(x,y) is bounded in some region about the point (x_0, y_0) and if f(x,y) satisfies the Lipschitz condition, *viz.*,

$$|f(x, y) - f(x, \bar{y})| \leq k|y - \bar{y}|$$

where k is a constant.

Here the same problem in Example **6.3.1** again solved using Picard's method in Example 6.4.1 below.

Example 6.4.1:

The output of an electric power plant for the orbiting space station uses photovoltaic cells to store energy in batteries. The charging system depends on the differential equation $\frac{dv}{dt} + 2v = 0$, with initial conditions $t_0 = 0$ & $v_0 = 1$.

Determine the value of v at t=0.2s using Picard's method correct to four decimal points.

Solution:

Here, $\frac{dv}{dt} = -2v = f(t,v)$

The initial conditions are given as $t_0 = 0$ & $v_0 = 1$.

We start with $v^{(0)} = 1$ and obtain,

$$v^{(1)} = 1 + \int_0^t (-2 * 1)dt$$

$$= 1 - 2\int_0^t dt$$

$$= 1 - 2t$$

Then the second approximation is,

$$v^{(2)} = 1 + \int_0^t (1 - 2t)dt$$

$$= 1 + \left[t - 2 * \frac{t^2}{2} \right]_0^t$$

$$= 1 + [t - 2t^2]_0^t$$

$$= 1 + t - 2t^2$$

It is seen that the integrations might become more and more difficult as we proceed to higher approximations. Hence this method is not recommended by the authors.

Euler's Method

Euler's method is a very well known method in which we obtain the solution as tabular values. For solution this method uses a simplest extrapolation technique.

Again here we assume a differential equation as in equation **(6.5.1)** and with initial conditions in equation **(6.5.1)** as follows:

$$\frac{dy}{dx} = f(x,y) \tag{6.5.1}$$

$$y = y_1 \text{ at } x = x_1 \tag{6.5.2}$$

From the geometrical explanation of Euler's method is illustrated in Fig. **(6.2)**, which shows that the Euler's method is basically a technique of developing a piecewise linear approximation of the solution [5].

At point (x_1, y_1) of the curve in Fig. **(6.2)**, the slope is obtained from **(6.2)** as

$$\frac{dy}{dx}(x_1, y_1) = f(x_1, y_1) \tag{6.5.3}$$

The next approximate solution y_2 can be obtained by means of extrapolation as follows :

$$y(x_1 + h) = y_2 = y_1 + hf(x_1, y_1) \tag{6.5.4}$$

Similarly, further approximations $y_3, y_4, \ldots\ldots$ can also be obtained from this method. In general the $(i + 1)^{th}$ approximation can be obtained as follows:

$$y_{i+1} = y_i + hf(x_i, y_i) \tag{6.5.5}$$

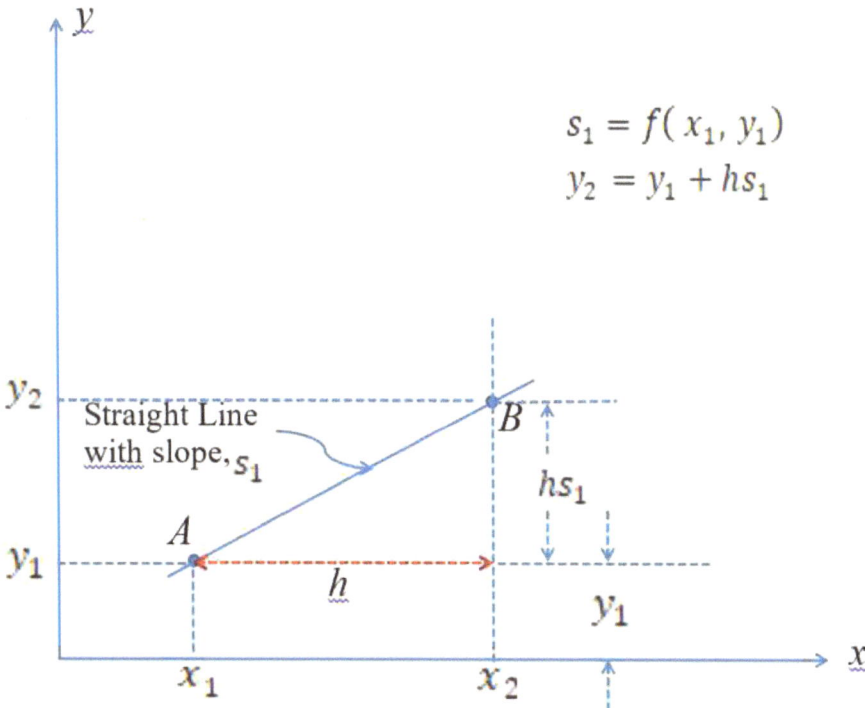

Fig. (6.2). Graphical representation of Euler's method.

An computer algorithm is provided here to ease the programming for the reader.

Algorithm 6.5.1: Steps of Heun's method to solve ODEs

1. Start

2. Define function to calculate slope $f(x, y)$

3. Declare variables

4. Input the value of variable

5. Find slope (say m_1) using initial values of x & y i.e $\frac{dy}{dx}(x_1, y_1) = f(x_1, y_1)$

6. Find new y , $y = y_1 + h * m_1$

7. Increase value of x , i.e $x_2 = x_1 + h$

8. Find new slope (say m_2) using x_2 and y

9. Set $m_1 = m_2$

10. Again find new y , and assign y_1 with y

11. Repeat from step 6 till two consecutive y are equal

12. Now, increase x and repeat from step 5 till $x = x_n$

13. Print x and corresponding y

14. Stop

Process of Euler's method to solve ODEs is described in Fig. (6.3).

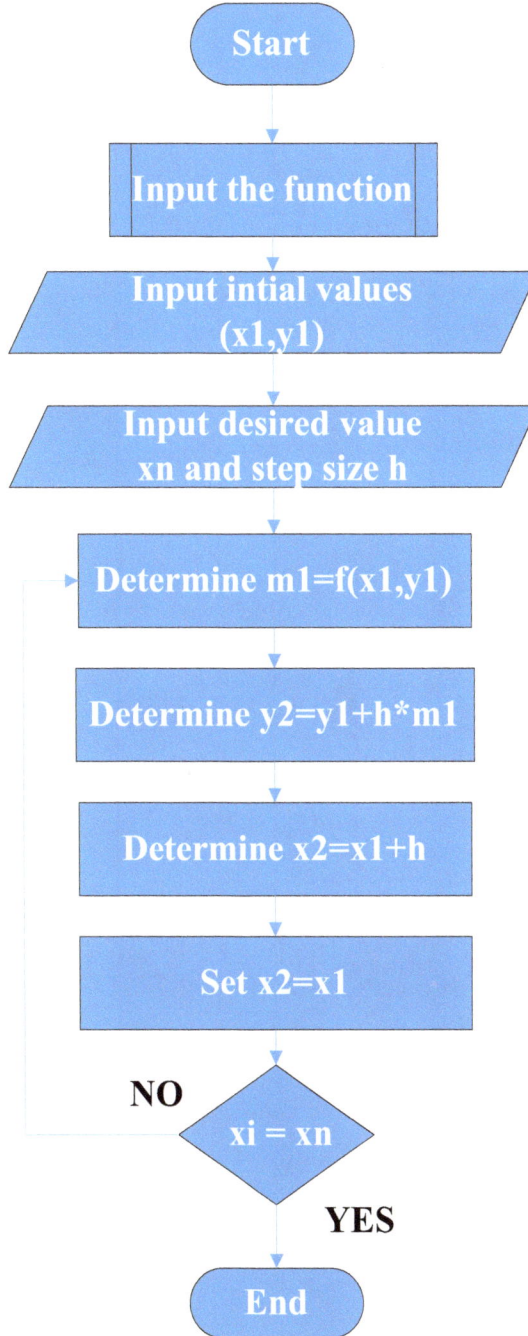

Fig.(6.3). Process of Euler's method to solve ODEs.

We solve another problem along with problem in example **6.3.1** again using Euler's method as follows.

Example 6.5.1

The output of an electric power plant for the orbiting space station uses photovoltaic cells to store energy in batteries. The charging system depends on the differential equation $\frac{dv}{dt} + 2v = 0$, with initial conditions $t_0 = 0$ & $v_0 = 1$.

Determine the value of v at t=0.2s using Euler's method.

Solution:

Here, $\frac{dv}{dt} = -2v = f(t, v)$

The initial conditions are given as $t_0 = 0$ & $v_0 = 1$.

For Euler's method to obtain reasonable accuracy the step size h should be as small as possible. But for simplicity we assume step size to be h=0.1.

Users are advised to take a smaller value of h.

Now,

$$f(t_0, v_0) = -2$$

The Euler's formula is:

$$v_{i+1} = v_i + hf(t_i, v_i) \tag{6.5.6}$$

Successive application of equation **(6.5.6)** with h=0.1 gives

$$v(0.1) = 1 + 0.1 * (-2) = 0.8$$

$$v(0.2) = 0.8 + 0.1 * (-1.6) = 0.64$$

Hence, the voltage v at t=0.2s is found to be 0.64.

Example 6.5.2

The temperature radiation of a ball in air at ambient temperature 300K can be described by the differential equation

$$\frac{d\emptyset}{dt} = -2.2067 \times 10^{-12}(\emptyset^4 - 81 \times 10^8)$$

Using Euler's method find the temperature of the ball at $t = 480$ seconds where $\emptyset$ is in K and t in second. It is assumed that the initial temperature of the ball is 1200K.

Solution:

Let, $f(t, \emptyset) = -2.2067 \times 10^{-12}(\emptyset^4 - 81 \times 10^8)$

And the step size is, $h = 60$

The Euler's formula is:

$\emptyset_{i+1} = \emptyset_i + hf(t_i, \emptyset_i)$

At $i = 0, t_0 = 0, \emptyset_0 = 1200K$

$$\begin{aligned}
\emptyset_1 &= \emptyset_0 + hf(t_0, \emptyset_0) \\
&= 1200 + 60\, f(0,1200) \\
&= 1200 + 60\, \{-2.2067 \\
&\qquad\qquad \times 10^{-12}(1200^4 - 81 \times 10^8)\} \\
&= 926.523669
\end{aligned}$$

At $i = 1, t_1 = 60, \emptyset_1 = 926.523669K$

$$\begin{aligned}
\emptyset_2 &= \emptyset_1 + hf(t_1, \emptyset_1) \\
&= 106.104 + 60\, f(60, 926.523669)
\end{aligned}$$

$$= 106.104 + 60 \{-2.2067$$
$$\times 10^{-12}(926.523669^4 - 81$$
$$\times 10^8)\}$$
$$= 830.0251561$$

Similarly,

$$\emptyset_3 = \emptyset_2 + hf(t_2, \emptyset_2) = 768.2542279$$
$$\emptyset_4 = \emptyset_3 + hf(t_3, \emptyset_3) = 723.2040204$$
$$\emptyset_5 = \emptyset_4 + hf(t_4, \emptyset_4) = 688.0573584$$
$$\emptyset_6 = \emptyset_5 + hf(t_5, \emptyset_5) = 659.4546511$$
$$\emptyset_7 = \emptyset_6 + hf(t_6, \emptyset_6) = 635.4871094$$
$$\emptyset_8 = \emptyset_7 + hf(t_7, \emptyset_7) = 614.9661409$$

Hence, $\emptyset(480) = \emptyset_8 = 614.9661409K$

6.6. Runge-Kutta Methods

The most popular and accurate numerical techniques used in finding approximate solution of a first-order initial value problem is the **Runge-Kutta methods.** Like Euler's method, these methods also calculate the slopes of solution curve in some discrete points of the independent variable and use the slopes to find the numerical solution of the ODE. There are Runge-Kutta methods of different orders depending on the slopes used in the calculation process. This section will demonstrate some of the Runge-Kutta methods with suitable numerical example.

6.6.1. Heun's Method

One of the second order R-K methods is Heun's method. As already mentioned in the previous section, one of the major disadvantages of Euler's method is its slowness. Also to obtain reasonable accuracy we should choose a small step size. Because of this restriction this method is not appropriate for practical use. Heun's method uses the modification of Euler's formula to improve the result. This formula uses the average of two adjacent slopes to obtain the next solution and gives a more accurate result. Result obtained from conventional Euler's method is significantly improved by this correction [6]. The solution process of the method is discussed below. Consider the following first order differential equation with given initial conditions.

$$\frac{dy}{dx} = f(x, y); \ y(x_1) = y_1$$

The method can be geometrically explained by the Fig. (6.4).

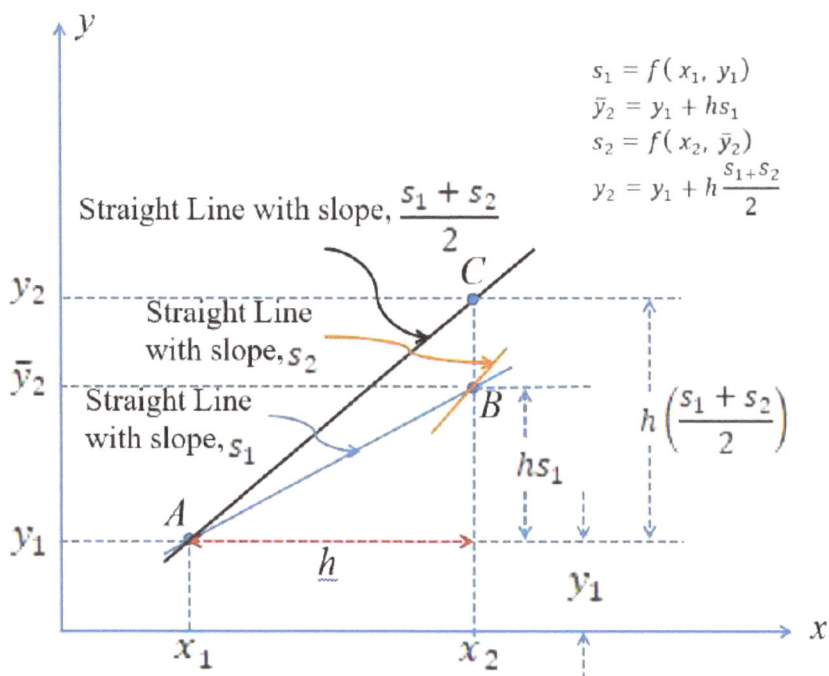

Fig. (6.4). Graphical explanation of Heun's method.

Here initial condition is the initial point of the solution curve as indicated by $A(x_1, y_1)$ in the figure. The slope $s_1 = f(x_1, y_1)$ is obtained from the above ODE using the initial values given. To obtain another slope draw a straight line from A with the previous slope s_1 and indicate the point of intersection with the vertical line at $x = x_2 = x_1 + h$. If the intersection point is $B(x_1 + h, \bar{y}_2)$ the second slope of the solution curve can be obtained as $s_2 = f(x_2, \bar{y}_2)$. Hence average of these two slopes is obtained as $s = \frac{(s_1+s_2)}{2}$ to use it for the final expression of next solution.

To obtain the next solution draw a straight line from A with the obtained average slope s and indicate the point of intersection again with the vertical line at $x = x_2 = x_1 + h$. The vertical coordinate of this intersection point is the final result of the next solution which is calculated as:

$$y_2 = y_1 + h\frac{(s_1 + s_2)}{2} \tag{6.6.1}$$

Similarly the 3rd solution at $x = x_3 = x_2 + h = x_1 + 2h$

$$y_3 = y_2 + h\frac{(s_2 + s_3)}{2}$$

Finally the nth solution at $x = x_n = x_{n-1} + h = x_1 + (n-1)h$

$$y_n = y_{n-1} + h\frac{(s_{n-1} + s_n)}{2}$$

Thus starting from i^{th} position the next solution formula can be generalized as:

$$y_{i+1} = y_i + h\frac{(s_i + s_{i+1})}{2} \tag{6.6.2}$$

Where, the slopes at two consecutive positions are obtained as $s_i = f(x_i, y_i)$ and $s_{i+1} = f(x_{i+1}, y_i + s_i h)$ respectively.

A computer algorithm to implement the above method is given below.

Algorithm 6.6.1: Steps of Heun's method to solve ODEs

1. Start
2. Input the function
3. Enter the values of initial points x_1, y_1.
4. Enter the point x_n ,where the solution is required.
5. Enter the step size h.
6. While $x_1 < x_n$ do the following steps.

7. Compute $s_1 = f(x_1, y_1)$
8. Compute $x_2 = x_1 + h$
9. Compute $y_2 = y_1 + h s_1$
10. Compute $s_2 = f(x_2, y_2)$
11. Compute $y_2 = y_1 + h \frac{s_1 + s_2}{2}$
12. Set $x_1 = x_2$ & Set $y_1 = y_2$
13. Return to step 5 again.
14. Stop

Process of Heun's method to solve ODEs is shown in Fig. (6.5).

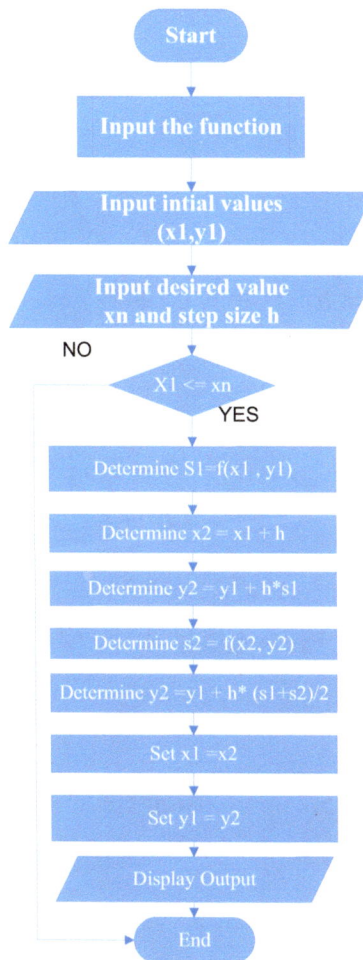

Fig. (6.5). Process of Heun's method to solve ODEs.

Problem statements of the Example 6.5.1 and Example 6.5.2 are solved again using Heun's method in the following Example 6.6.1 and Example 6.6.2.

Example 6.6.1

The output of an electric power plant for the orbiting space station uses photovoltaic cells to store energy in batteries. The charging system depends on the differential equation $\frac{dv}{dt} + 2v = 0$, with initial conditions $t_0 = 0$ & $v_0 = 1$.

Determine the value of v at $t=0.2$s using Heun's method.

Solution:

Here, $\frac{dv}{dt} = -2v = f(t, v)$

The initial conditions are given as $t_0 = 0$ & $v_0 = 1$.

We choose h=0.1

$$k_1 = hf_0 = 0.1 \times (-2) = -0.2$$
$$k_2 = hf(t_0 + h, v_0 + k_1) = -0.16$$
$$v(0.1) = v_1 = v_0 + 0.5 \times (k_1 + k_2) = 0.82$$

Similarly we get,

$$f(t_1, v_1) = -1.64$$
$$k_1 = -0.164$$
$$k_2 = -0.1312$$
$$v_2 = v(0.2) = 0.6724$$

This is the required answer.

Example 6.6.2

We will solve the example **6.5.2** again using Heun's method

Let, $f(t, \emptyset) = -2.2067 \times 10^{-12}(\emptyset^4 - 81 \times 10^8)$

And the step size is, $h = 60$

The Heun's formula is:

$$\emptyset_{i+1} = \emptyset_i + h\frac{1}{2}(s_1 + s_2)$$

$$s_1 = f(t_i, \emptyset_i)$$

$$s_2 = f(t_i + h, \emptyset_i + s_1 h)$$

At $i = 0, t_0 = 0, \emptyset_0 = 1200K$

$$s_1 = f(t_0, \emptyset_0) = -4.557939$$

$$s_2 = f(t_0 + h, \emptyset_0 + s_1 h) = f(60, 926.523669)$$

$$= -1.608309$$

Then, $\emptyset_1 = \emptyset_0 + h\dfrac{1}{2}(s_1 + s_2)$

$$= 1200 + 60\frac{1}{2}\{(-4.5579) + (-1.608309)\}$$

$$= 1015.012578$$

At $i = 1, t_1 = 60, \emptyset_1 = 1015.012578\ K$

$$s_1 = f(t_1, \emptyset_1) = -2.324353$$

$$s_2 = f(t_1 + h, \emptyset_1 + s_1 h) = -1.278916$$

Then, $\emptyset_2 = \emptyset_1 + h\dfrac{1}{2}(s_1 + s_2)$

$$= 655.16 + 240\frac{1}{2}\{(-2.324353) + (-1.278916)\}$$

$$= 906.914500$$

Similarly,

$$\emptyset_3 = \emptyset_2 + h\frac{1}{2}(s_1 + s_2) = 833.501701$$

$$\emptyset_4 = \emptyset_3 + h\frac{1}{2}(s_1 + s_2) = 779.269789$$

$$\emptyset_5 = \emptyset_4 + h\frac{1}{2}(s_1 + s_2) = 736.972879$$

$$\emptyset_6 = \emptyset_5 + h\frac{1}{2}(s_1 + s_2) = 702.713601$$

$$\emptyset_7 = \emptyset_6 + h\frac{1}{2}(s_1 + s_2) = 674.183138$$

$$\emptyset_8 = \emptyset_7 + h\frac{1}{2}(s_1 + s_2) = 649.913562$$

hence, $\qquad \emptyset(480) = \emptyset_8 = 649.913562K$

Polygon Method

Another improved second order R-K method called Polygon method is in practice instead of Heun's method [5, 6]. Unlike Heun's method, Polygon method calculates the slope s_2 in the middle of the solution span. Instead of averaging of the two adjacent slopes it uses only the second slope to obtain the next solution and gives a more accurate result. The solution process of the method is discussed below.

Consider the following first order differential equation with given initial conditions

$$\frac{dy}{dx} = f(x, y); \; y(x_1) = y_1$$

The method can be geometrically explained by the Fig. (**6.6**).

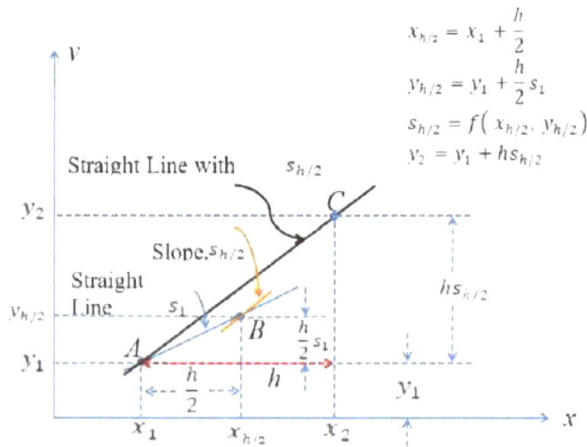

Fig. (**6.6**). Illustration of polygon method.

Here initial condition is the initial point of the solution curve as indicated by $A(x_1, y_1)$ in the figure. The slope $s_1 = f(x_1, y_1)$ is obtained from the above ODE using the initial values given. To obtain another slope draw a straight line from A with the previous slope s_1 and indicate the point of intersection with the vertical line erected at $x = x_{h/2} = x_1 + \dfrac{h}{2}$. Vertical coordinate, $y_{h/2}$ at this position can be calculated as $y_{h/2} = y_1 + \dfrac{h}{2}s_1$. Now the intersection point is $B(x_{h/2}, y_{h/2})$ and using this coordinate the second slope of the solution curve can be obtained as $s_{h/2} = f(x_{h/2}, y_{h/2})$. Hence this second slope is used for the final expression of next solution. To obtain the next solution draw again a straight line from A with the slope s_2 and indicate the point of intersection with the vertical line erected at $x = x_2 = x_1 + h$. The vertical coordinate of this intersection point is the final result of the next solution which is calculated as:

$$y_2 = y_1 + hs_{h/2} \qquad (6.6.3)$$

Similarly the 3rd solution at $x = x_3 = x_2 + h = x_1 + 2h$

$$y_3 = y_2 + hs_{3h/2}$$

Finally the nth solution at $x = x_n = x_{n-1} + h = x_1 + (n-1)h$

$$y_n = y_{n-1} + hs_{nh/2}$$

Thus starting from i^{th} position the next solution formula can be generalized as:

$$y_{i+1} = y_i + hs_{(i+1)h/2} \qquad (6.6.4)$$

with
$$s_{(i+1)h/2} = f\left(x_i + \frac{h}{2}, y_i + s_i\frac{h}{2}\right)$$

and
$$s_i = f(x_i, y_i)$$

Example 6.6.3

We will solve the example **6.5.2** again using Polygon method

Let, $f(t, \emptyset) = -2.2067 \times 10^{-12}(\emptyset^4 - 81 \times 10^8)$

And the step size is, $h = 60$

The Polygon formula is:

$$\emptyset_{i+1} = \emptyset_i + hf(t_i + \frac{h}{2}, \emptyset_i + s_i\frac{h}{2})$$

$$s_i = f(t_i, \emptyset_i)$$

At $i = 0, t_0 = 0, \emptyset_0 = 1200K$

$\quad s_0 = f(t_0, \emptyset_0) = -4.557939$

Then
$$\emptyset_1 = \emptyset_0 + hf(t_0 + \frac{h}{2}, \emptyset_0 + s_0\frac{h}{2})$$

$$= \emptyset_0 + hf(30, 1063.26183)$$

$$= 1200 + 60(-2.802483)$$

$$= 1031.850996$$

Similarly,

$$\emptyset_2 = \emptyset_1 + hf\left(t_1 + \frac{h}{2}, \emptyset_1 + s_1\frac{h}{2}\right) = 921.709303$$

$$\emptyset_3 = \emptyset_2 + hf\left(t_2 + \frac{h}{2}, \emptyset_2 + s_2\frac{h}{2}\right) = 845.359369$$

$$\emptyset_4 = \emptyset_3 + hf(t_3 + \frac{h}{2}, \emptyset_3 + s_3\frac{h}{2}) = 788.847498$$

$$\emptyset_5 = \emptyset_4 + hf(t_4 + \frac{h}{2}, \emptyset_4 + s_4\frac{h}{2}) = 744.869821$$

$$\emptyset_6 = \emptyset_5 + hf(t_5 + \frac{h}{2}, \emptyset_5 + s_5\frac{h}{2}) = 709.356606$$

$$\emptyset_7 = \emptyset_6 + hf(t_6 + \frac{h}{2}, \emptyset_6 + s_6\frac{h}{2}) = 679.868897$$

$$\emptyset_8 = \emptyset_7 + hf(t_7 + \frac{h}{2}, \emptyset_7 + s_7\frac{h}{2}) = 654.851416$$

Hence, $\emptyset(480) = \emptyset_8 = 654.851416K$

6.6.3. Runge-Kutta Fourth Order Formula

In the last section it was shown that using two estimates of the slope (*i.e.*, in Second Order Runge Kutta) gave an approximation with greater accuracy than using just a single slope (*i.e.*, First Order Runge Kutta; basically Euler's method is said to be a first-order Runge-Kutta method). It seems reasonable, then, to assume that using even more estimates of the slope would result in even more accuracy. It turns out this is true, and leads to higher-order Runge-Kutta methods. Third order methods can be developed (but are not discussed here). Instead we will restrict ourselves to the much more commonly used Fourth Order Runge-Kutta technique, which uses four approximations to the slope.

Again consider the following first order differential equation with given initial conditions

$$\frac{dy}{dx} = f(x, y); \; y(x_1) = y_1$$

For more error improvements Taylor series approximation of the solution is truncated after taking five terms up-to the h^4 term as follows:

$$y_2 = y_1 + hy_1' + \frac{h^2}{2!} y_1'' + \frac{h^3}{3!} y_1''' + \frac{h^4}{4!} y_1^{iv}$$

To avoid the need to calculating the derivatives of f leads to the Runge-Kutta fourth order formulae. This formula is extensively used in practice.

This method is geometrically illustrated in Fig. **(6.7)** as follows:

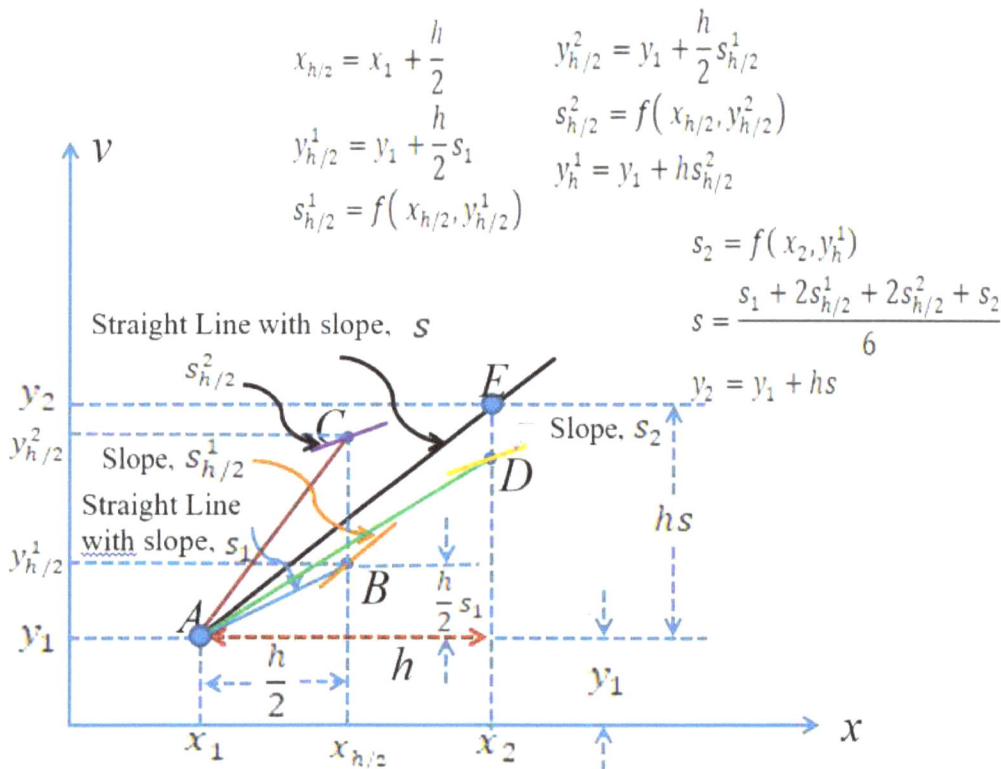

$$x_{h/2} = x_1 + \frac{h}{2}$$

$$y_{h/2}^1 = y_1 + \frac{h}{2} s_1$$

$$s_{h/2}^1 = f\left(x_{h/2}, y_{h/2}^1\right)$$

$$y_{h/2}^2 = y_1 + \frac{h}{2} s_{h/2}^1$$

$$s_{h/2}^2 = f\left(x_{h/2}, y_{h/2}^2\right)$$

$$y_h^1 = y_1 + h s_{h/2}^2$$

$$s_2 = f\left(x_2, y_h^1\right)$$

$$s = \frac{s_1 + 2 s_{h/2}^1 + 2 s_{h/2}^2 + s_2}{6}$$

$$y_2 = y_1 + hs$$

Fig. (6.7). Fourth order Runge-Kutta method graphical illustration.

This method uses the average of four slopes to obtain the next solution. Here, the initial condition is the initial point of the solution curve as indicated by $A(x_1, y_1)$ in the figure. The slope $s_1 = f(x_1, y_1)$ is obtained from the above ODE using the initial values given. To obtain the second slope, draw a straight line from A with the previous slope s_1 and indicate the point of intersection with the vertical line erected at $x = x_{h/2} = x_1 + \frac{h}{2}$. Vertical coordinate, $y_{h/2}^1$ at this position can be calculated as $y_{h/2}^1 = y_1 + \frac{h}{2}s_1$. Now the intersection point is $B\left(x_{h/2}, y_{h/2}^1\right)$ and using this coordinate, the second slope of the solution curve can be obtained as $s_{h/2}^1 = f\left(x_{h/2}, y_{h/2}^1\right)$. To obtain the third slope, draw a straight line from A with the second slope $s_{h/2}^1$ and indicate the point of intersection again with the vertical line erected at $x = x_{h/2} = x_1 + \frac{h}{2}$. Vertical coordinate, $y_{h/2}^2$ at this position can be calculated as $y_{h/2}^2 = y_1 + \frac{h}{2}s_{h/2}^1$. Now the intersection point is $C\left(x_{h/2}, y_{h/2}^2\right)$ and using this coordinate, the third slope of the solution curve can be obtained as $s_{h/2}^2 = f\left(x_{h/2}, y_{h/2}^2\right)$. To obtain the fourth slope draw a straight line from A with the third slope $s_{h/2}^2$ and indicate the point of intersection with the vertical line erected at $x = x_2 = x_1 + h$. Vertical coordinate, y_h at this position can be calculated as $y_h = y_1 + hs_{h/2}^2$. Now the intersection point is $D(x_2, y_h)$ and using this coordinate, the fourth slope of the solution curve can be obtained as $s_2 = f(x_2, y_h)$. Hence, weighted average of these four slopes is obtained as

$$s = \frac{(s_1 + 2s_{h/2}^1 + 2s_{h/2}^2 + s_2)}{6}$$

Hence, this slope s is used for the final expression of the next solution. To obtain the next solution draw again a straight line from A with the slope s and indicate the point of intersection with the vertical line erected at $x = x_2 = x_1 + h$. The vertical coordinate of this intersection point is the final result of the next solution which is calculated as:

$$y_2 = y_1 + hs \tag{6.6.5}$$

where

$$s = \frac{s_1 + 2s_{h/2}^1 + 2s_{h/2}^2 + s_2}{6}$$

$$x_{h/2} = x_1 + \frac{h}{2}$$

$$y_{h/2}^1 = y_1 + \frac{h}{2} s_1$$

$$s_{h/2}^1 = f\left(x_{h/2}, y_{h/2}^1\right)$$

$$y_{h/2}^2 = y_1 + \frac{h}{2} s_{h/2}^1$$

$$s_{h/2}^2 = f\left(x_{h/2}, y_{h/2}^2\right)$$

$$y_h^1 = y_1 + h s_{h/2}^2$$

$$s_2 = f\left(x_2, y_h^1\right)$$

Thus starting from i^{th} position the next solution formula can be generalized as:

$$y_{i+1} = y_i + h_i s \qquad\qquad (6.6.6)$$

with

$$s = \frac{\left(s_i + 2s_{h/2}^i + 2s_{h/2}^{i+1} + s_{i+1}\right)}{6}$$

and

$$s_i = f(x_i, y_i)$$

$$y_{h/2}^i = y_i + \frac{h}{2} s_i$$

$$s_{h/2}^i = f\left(x_{h/2}, y_{h/2}^i\right)$$

$$y_{h/2}^{i+1} = y_i + \frac{h}{2} s_{h/2}^i$$

$$s_{h/2}^{i+1} = f\left(x_{h/2}, y_{h/2}^{i+1}\right)$$

$$y_h^i = y_i + h_i s_{h/2}^{i+1}$$

$$s_{i+1} = f\left(x_{i+1}, y_h^i\right)$$

An algorithm for the fourth order Runge-Kutta method is given below. Observe that we have to find $f(x, y)$ four times to calculate each value of the solution.

Algorithm 6.6.2: Steps of R-K Fourth Order Method to Solve ODEs

1. Start

2. Input the function

3. Enter the values of initial points x_1, y_1.

4. Enter the point x_n ,where the solution is required.

5. Enter the step size h.

6. While $x_1 < x_n$ do the following steps.

7. Compute $s_1 = f(x_1, y_1)$

8. Compute $x_2 = x_1 + \dfrac{h}{2}$

9. Compute $y_2 = y_1 + \dfrac{hs_1}{2}$

10. Compute $s_2 = f(x_2, y_2)$

11. Compute $y_2 = y_1 + h\dfrac{s_2}{2}$

12. Compute $s_3 = f(x_2, y_2)$

13. Compute $x_2 = x_1 + h$

14. Compute $y_2 = y_1 + hs_3$

15. Compute $s_4 = f(x_2, y_2)$

16. Compute $s = (s_1 + 2s_2 + 2s_3 + s_4)$

17. Compute $y_2 = y_1 + h\dfrac{s}{6}$

18. Set $x_1 = x_2$

19. Set $y_1 = y_2$

20. Return to step 5 again.

21. Stop

Process of R-K fourth order method to solve ODEs is shown in Fig.(6.8).

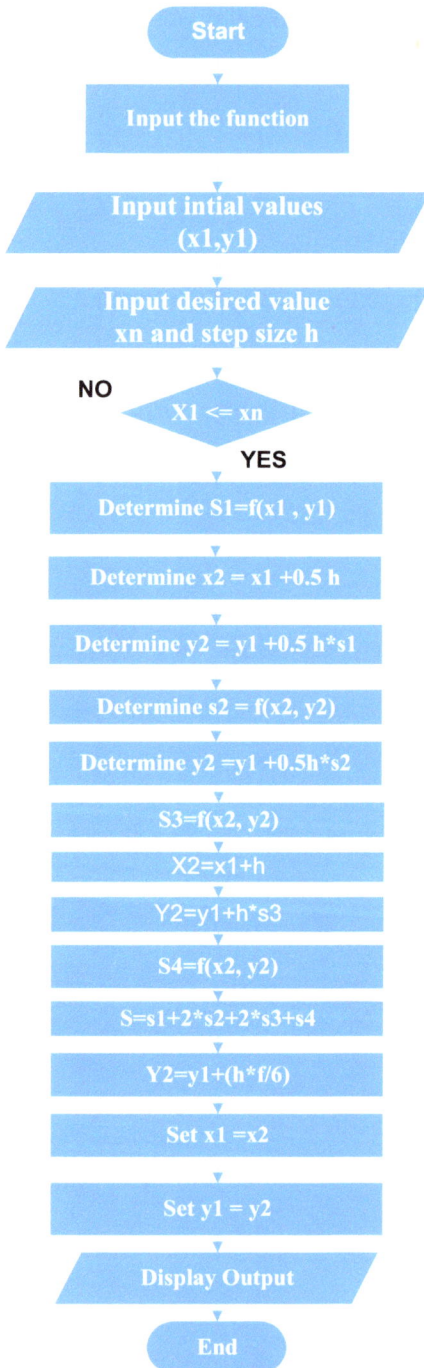

Fig.(6.8). Process of R-K fourth order method to solve ODEs.

Example 6.6.4.

The output of an electric power plant for the orbiting space station uses photovoltaic cells to store energy in batteries. The charging system depends on the differential equation $\frac{dv}{dt} + 2v = 0$, with initial conditions $t_0 = 0$ & $v_0 = 1$.

Determine the value of v at $t=0.2$ using Runge-Kutta Fourth Order Formula.

Solution:

Here, $\frac{dv}{dt} = -2v = f(t, v)$

The initial conditions are given as $t_0 = 0$ & $v_0 = 1$.

We choose h=0.1

The Runge-Kutta Fourth Order Formula is:

$$v_{i+1} = v_i + \frac{h}{6}(s_1 + 2s_2 + 2s_3 + s_4)$$

Now, $s_1 = hf_0 = -0.2$

$s_2 = hf\left(t_0 + \frac{h}{2}, v_0 + \frac{s_1}{2}\right) = -0.18$

$s_3 = hf\left(t_0 + \frac{h}{2}, v_0 + \frac{s_2}{2}\right) = -0.182$

$s_4 = hf(t_0 + h, v_0 + s_3) = -0.1636$

$v_1 = v(0.1) = v_0 + \frac{1}{6}(s_1 + 2s_2 + 2s_3 + s_4) = 0.82$

Again, $s_1 = hf(t_1, v_1) = -0.164$

$s_2 = hf\left(t_1 + \frac{h}{2}, v_1 + \frac{s_1}{2}\right) = -0.1476$

$s_3 = hf\left(t_1 + \frac{h}{2}, v_1 + \frac{s_2}{2}\right) = -0.1492$

$s_4 = hf(t_1 + h, v_1 + s_3) = -0.134152$

$v_2 = v(0.2) = v_1 + \frac{1}{6}(s_1 + 2s_2 + 2s_3 + s_4) = 0.671$

Example 6.6.5:

We will solve the example **6.5.2** again using fourth order Runge-Kutta method.

Let, $f(t, \emptyset) = -2.2067 \times 10^{-12}(\emptyset^4 - 81 \times 10^8)$

And the step size is, $h = 60$

The Runge-Kutta Fourth Order Formula is:

$$\emptyset_{i+1} = \emptyset_i + \frac{h}{6}(s_1 + 2s_2 + 2s_3 + s_4)$$

$$s_1 = f(t_i, \emptyset_i)$$

$$s_2 = f(t_i + \frac{h}{2}, \emptyset_i + s_1 \frac{h}{2})$$

$$s_3 = f\left(t_i + \frac{h}{2}, \emptyset_i + s_2 \frac{h}{2}\right)$$

$$s_4 = f(t_i + h, \emptyset_i + s_3 h)$$

At $i = 0, t_0 = 0, \emptyset_0 = 1200K$

$$s_1 = f(t_0, \emptyset_0) = -4.557939$$

$$s_2 = f\left(t_0 + \frac{h}{2}, \emptyset_0 + s_1 \frac{h}{2}\right) = -2.802483$$

$$s_3 = f\left(t_0 + \frac{h}{2}, \emptyset_0 + s_2 \frac{h}{2}\right) = -3.404158$$

$$s_4 = f(t_0 + h, \emptyset_0 + s_3 h) = -2.151555$$

Then, $\emptyset_1 = \emptyset_0 + \frac{h}{6}(s_1 + 2s_2 + 2s_3 + s_4)$

$$= 1008.772235$$

Similarly,

$$\emptyset_2 = \emptyset_1 + hf\left(t_1 + \frac{h}{2}, \emptyset_1 + s_1 \frac{h}{2}\right) = 900.948793$$

$$\emptyset_3 = \emptyset_2 + hf\left(t_2 + \frac{h}{2}, \emptyset_2 + s_2\frac{h}{2}\right) = 828.442701$$

$$\emptyset_4 = \emptyset_3 + hf\left(t_3 + \frac{h}{2}, \emptyset_3 + s_3\frac{h}{2}\right) = 775.020473$$

$$\emptyset_5 = \emptyset_4 + hf\left(t_4 + \frac{h}{2}, \emptyset_4 + s_4\frac{h}{2}\right) = 733.368233$$

$$\emptyset_6 = \emptyset_5 + hf\left(t_5 + \frac{h}{2}, \emptyset_5 + s_5\frac{h}{2}\right) = 699.615328$$

$$\emptyset_7 = \emptyset_6 + hf\left(t_6 + \frac{h}{2}, \emptyset_6 + s_6\frac{h}{2}\right) = 671.486188$$

$$\emptyset_8 = \emptyset_7 + hf\left(t_7 + \frac{h}{2}, \emptyset_7 + s_7\frac{h}{2}\right) = 647.539297$$

Hence, $\emptyset(480) = \emptyset_8 = 647.539297K$

Predictor-corrector Method

In Predictor-collector method, one uses an iterative formula together with another formula (the predictor) whose mission is to provide an intelligent first guess for the iterative method to use. The predictor formula will be explicit, or non-iterative. If the predictor formula is clever enough, then it will happen that just a single application of the iterative refinement (corrector formula) will be sufficient, and we won't have to get involved in a long convergence process. There are various predictor-corrector methods in practice. The predictor-corrector method presented here is of a class called *Milne's* predictor-corrector method. In its basic form, we assumed a quadratic solution curve for extrapolation and a linear derivative curve. The corrector uses Trapezoidal rule for integration (which is valid for a quadratic).

Now, consider the following differential equations of the type

$$\frac{dy}{dx} = f(x, y)$$

Here, initial condition is the initial point of the solution curve as indicated by $A(x_1, y_1)$ in the figure.

Then, next solution y_2 is obtained from Heun's method using the following formula

$$y_2 = y_1 + h\frac{(s_1 + s_2)}{2} \qquad (6.7.1)$$

Where $\qquad\qquad\qquad s_1 = f(x_1, y_1)$

And $\qquad\qquad\qquad s_2 = f(x_2, \bar{y}_2) = f(x_2, y_1 + hs_1)$

The next solution y_3 at x_3 can be found from two past information *i.e.* $A(x_1, y_1)$ and $B(x_2, y_2)$ by fitting a quadratic through them.

Consider the quadratic through $A(x_1, y_1)$ and $B(x_2, y_2)$ is obtained as follows:

$$y = P + Q(x - x_1) + R(x - x_1)(x - x_2) \qquad (6.7.2)$$
$$= P + Q(x - x_1) + R(x - x_1)(x - x_1 - h)$$

Obtaining the Coefficient, P: From the initial condition we have,

$$y = y_1 \quad \text{at} \quad x = x_1$$

Substituting these values in (6.7.2) gives

$$P = y_1 \qquad (6.7.3)$$

Using **(6.7.3)** expression **(6.7.2)** can be rewritten as:

$$y = y_1 + Q(x - x_1) + R(x - x_1)(x - x_2) \qquad (6.7.4)$$

Obtaining the Coefficient, Q: From the Heun's method we obtained,

$$y = y_2 \quad \text{at} \quad x = x_2 = x_1 + h$$

Substituting these known values in **(6.7.4)** gives

$$y_2 = y_1 + Q(x_2 - x_1) + R(x_2 - x_1)(x_2 - x_2)$$

After manipulation,

$$Q = \frac{y_2 - y_1}{h} \qquad (6.7.5)$$

Obtaining the Coefficient, R:

We have obtained the values of P and Q already and also from the given differential equation we can have,

$$\frac{dy}{dx}(x_1, y_1) = f(x_1, y_1)$$

And

$$\frac{dy}{dx}(x_2, y_2) = f(x_2, y_2) \tag{6.7.6}$$

Now the coefficients R can be determined by taking first derivative of **(6.7.2)** and then substituting the known values of Q and $\frac{dy}{dx}$.

$$\frac{dy}{dx} = Q + R(2x - x_1 - x_2)$$

$$\frac{dy}{dx}(x_2, y_2) = f(x_2, y_2) = \frac{y_2 - y_1}{h} + R(2x_2 - x_1 - x_2)$$

$$f(x_2, y_2) = \frac{y_2 - y_1}{h} + R(x_2 - x_1) \tag{6.7.7}$$

$$f(x_2, y_2) = \frac{y_2 - y_1}{h} + Rh$$

Therefore,

$$R = \frac{1}{h}f(x_2, y_2) - \frac{y_2 - y_1}{h^2} \tag{6.7.8}$$

Using these values of P, Q and R the value of y at x_3 may be obtained from equation **(6.10.1)** as

$$y_3^p = y_1 + \frac{y_2 - y_1}{h}(x_3 - x_1) + \left\{\frac{1}{h}f(x_2, y_2) - \frac{y_2 - y_1}{h^2}\right\}(x_3 - x_1)(x_3 - x_2)$$

$$= y_1 + \frac{y_2 - y_1}{h}(2h) + \left\{\frac{1}{h}f(x_2, y_2) - \frac{y_2 - y_1}{h^2}\right\}(2h)(h)$$

$$= y_1 + \frac{y_2 - y_1}{h}(2h) + 2hf(x_2, y_2) - \frac{y_2 - y_1}{h}(2h)$$

$$= y_1 + 2hf(x_2, y_2)$$

This is the predicted value of the next solution at x_3 . Now we need an improvement of this result by the following correction using Heun's method starting from $B(x_2, y_2)$.

$$y_3^c = y_2 + h \left(\frac{s_2 + s_3}{2} \right)$$

Where

$$s_2 = f(x_2, y_2)$$

And

$$s_3 = f\left(x_3, y_3^p\right)$$

Finally the formula can be written as follows.

$$y_3 = y_3^c = y_2 + \frac{h}{2} \{f(x_2, y_2) + f\left(x_3, y_3^p\right)\}$$

Similarly we can proceed for the fourth solution by the following expressions:

$$y_4^p = y_2 + 2hf(x_3, y_3)$$

$$y_4^c = y_3 + \frac{h}{2} \{f(x_3, y_3) + f\left(x_4, y_4^p\right)\}$$

Thus, generalized formula can be written as follows:

$$y_{i+2}^p = y_i + 2hf(x_{i+1}, y_{i+1}) \tag{6.7.9}$$

$$y_{i+2} = y_{i+2}^c = y_{i+1} + \frac{h}{2} \{f(x_{i+1}, y_{i+1}) + f\left(x_{i+2}, y_{i+2}^p\right)\} \tag{6.7.10}$$

As it uses the previously predicted information of the solution in the successive iterations predictor – corrector method exhibits better performance compared to others.

Let us solve the problem in Example **6.5.2** again using predictor-corrector method.

Example 6.7.1:

Let, $f(t, \emptyset) = -2.2067 \times 10^{-12}(\emptyset^4 - 81 \times 10^8)$

And the step size is, $h = 60$

The predictor-corrector formulae are:

$$\emptyset_{i+2}^p = \emptyset_i + 2hf(t_{i+1}, \emptyset_{i+1})$$

$$\emptyset_{i+2}^c = \emptyset_{i+1} + \frac{h}{2}[f(t_{i+1}, \emptyset_{i+1}) + f(t_{i+2}, \emptyset_{i+2}^p)]$$

Started with Heun's method,

$$\emptyset(0) = 1200, \emptyset(60) = 906.914500$$

Applying predictor-corrector formulae,

$$\emptyset_1 = 1200, \emptyset_2 = 906.914500, t_1 = 0, t_2 = 60$$

$$\emptyset_3^p = \emptyset_1 + 2hf(t_2, \emptyset_2)$$

$$= 1200 + 120f(60, 906.914500)$$

$$= 1023.006004$$

$$\emptyset_3^c = \emptyset_2 + \frac{h}{2}[f(t_2, \emptyset_2) + f(t_3, \emptyset_3^p)]$$

$$= 906.914500 + 30[f(60,906.914500) + f(180,1023.006004)]$$

$$= 790.695673$$

Similarly,

$$\emptyset_4^c = \emptyset_3 + \frac{h}{2}[f(t_3, \emptyset_3) + f(t_4, \emptyset_4^p)] = 738.014987$$

$$\emptyset_5^c = \emptyset_4 + \frac{h}{2}[f(t_4, \emptyset_4) + f(t_5, \emptyset_5^p)] = 702.215725$$

$$\emptyset_6^c = \emptyset_5 + \frac{h}{2}\left[f(t_5, \emptyset_5) + f\left(t_6, \emptyset_6^p\right)\right] = 673.385225$$

$$\emptyset_7^c = \emptyset_6 + \frac{h}{2}\left[f(t_6, \emptyset_6) + f\left(t_7, \emptyset_7^p\right)\right] = 649.034809$$

$$\emptyset_8^c = \emptyset_7 + \frac{h}{2}\left[f(t_7, \emptyset_7) + f\left(t_8, \emptyset_8^p\right)\right] = 628.027653$$

$$\emptyset_9^c = \emptyset_8 + \frac{h}{2}\left[f(t_8, \emptyset_8) + f\left(t_9, \emptyset_9^p\right)\right] = 609.636280$$

More sophisticated predictor-corrector formulae are used in practice which uses fourth degree solution curve for extrapolation and a cubic derivative curve. The corrector uses Simpson's rule for integration (which is valid for a cubic). The equations are:

$$y_{i+1}^p = y_{i-3} + \frac{4h}{3}(2y_i' - y_{i-1}' + 2y_{i-2}') \tag{6.10.15}$$

$$y_{i+1}^c = y_{i-1} + \frac{h}{3}(y_{i-1}' - 4y_i' + y_{i+1}') \tag{6.10.16}$$

6.8. Solving Simultaneous First Order ODEs and Higher Order ODEs

In the previous sections, so far, we considered only single first order differential equation. The process is equally applicable for the multiple first order ODEs. Using this concept higher order ODEs can also be solved as they can be represented by some of the simultaneous first order ODEs. Any of the previous solution techniques may be used for this purpose; however, Heun's method is discussed here.

6.8.1 Solving Simultaneous First Order ODEs

Let consider the following two equations:

$$\frac{dy}{dx} = f(x, y, w); \text{ where, } y(x_1) = y_1. \tag{6.8.1}$$

$$\frac{dw}{dx} = g(x, y, w); \text{ where, } w(x_1) = w_1. \tag{6.8.2}$$

Heun's method gives the following two generalized solution equations for y and w:

$$y_{i+1} = y_i + h\frac{(s_i + s_{i+1})}{2} \tag{6.8.3}$$

$$w_{i+1} = w_i + h\frac{(p_i + p_{i+1})}{2} \tag{6.8.4}$$

where,

$$s_i = f(x_i, y_i, w_i)$$

$$s_{i+1} = f(x_i + h, y_i + hs_i, w_i + hp_i)$$

$$p_i = g(x_i, y_i, w_i)$$

$$p_{i+1} = g(x_i + h, y_i + hs_i, w_i + hp_i)$$

6.8.2. Solving Higher Order ODEs

Any N^{th} order ODE can be represented by N number of simultaneous first order ODEs. For example, a second order ODE can be represented by two simultaneous first order ODEs. Thus, the higher order ODEs can be solved using the process of solving simultaneous first order ODEs as discussed above.

Let consider the following second order differential equation:

$$\frac{d^2y}{dx^2} = g\left(x, y, \frac{dy}{dx}\right) \tag{6.8.5}$$

With initial conditions given as follows:

$$y(x_1) = y_1 \quad \text{and} \quad \left(\frac{dy}{dx}\right)_{x_1} = y_1' \tag{6.8.6}$$

We can express (6.8.5) by the following two simultaneous first order ODEs:

$$\frac{dy}{dx} = w \tag{6.8.7}$$

$$\text{and} \quad \frac{dw}{dx} = g(x, y, w) \tag{6.8.8}$$

With newly defined initial conditions

$$y(x_1) = y_1 \quad \text{and} \quad w(x_1) = w_1 = y_1'$$

We can rewrite (6.8.7) as

$$\frac{dy}{dx} = f(x, y, w) \tag{6.8.9}$$

Now, we have exactly similar ODEs in **(6.8.9)** and **(6.8.8)** as in **(6.8.1)** and **(6.8.2)**.

These can be solved following the process discussed in section 6.8.1. Finally, the required equations of Heun's method corresponding to solve this second order ODE are given below.

$$y_{i+1} = y_i + h\frac{(s_i+s_{i+1})}{2} \tag{6.8.10}$$

$$w_{i+1} = w_i + h\frac{(p_i+p_{i+1})}{2} \tag{6.8.11}$$

where,

$$s_i = f(x_i, y_i, w_i) = w_i$$

$$s_{i+1} = f(x_i + h, y_i + hs_i, w_i + hp_i) = w_i + hp_i$$

$$p_i = g(x_i, y_i, w_i)$$

$$p_{i+1} = g(x_i + h, y_i + hs_i, w_i + hp_i)$$

Now, we are concerned to get solutions of simultaneous differential equations of a practical problem in the following Example 6.8.1.

Example 6.8.1

The response of the following series RLC circuit can be represented by:

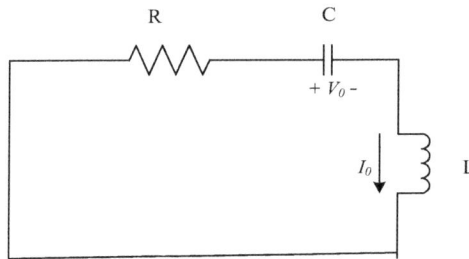

$$\frac{d^2i}{dt^2} + \frac{R}{L}\frac{di}{dt} + \frac{i}{LC} = 0$$

$$\frac{di}{dt} = -\frac{1}{L}(RI_0 + V_0)$$

where, R=10 Ω, L=2H and C=0.5F

Find the current of the circuit at $t = 0.5$s. Assume that the current through the inductor and the voltage across the capacitor at t=0s are i(0)=I_0=0A and v(0)=V_0= 10V respectively.

Solution:

$$\frac{d^2i}{dt^2} + \frac{R}{L}\frac{di}{dt} + \frac{i}{LC} = 0$$

Or,

$$\frac{d^2i}{dt^2} + 5\frac{di}{dt} + i = 0$$

Assuming

$$\frac{di}{dt} = z$$

then,

$$\frac{dz}{dt} = \frac{d^2i}{dt^2} = -5\frac{di}{dt} - i = -5z - i$$

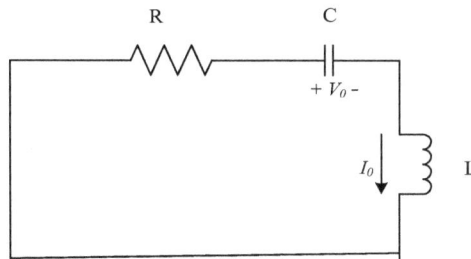

To Solve the Equations Following Initial Conditions to be Considered

$t_1 = 0, i_1 = 0, \frac{di(0)}{dt} = \mathbf{z(0)} = z_1 = -\frac{1}{2}(10 \times 0 + 10) = -5$

Let the step size be, $h = 0.25$

$i_2 = i_1 + \dfrac{h(s_1 + s_2)}{2}$

$s_1 = z_1 = -5$

$p_1 = -5z_1 - i_1 = 25$

$$s_2 = z_1 + hp_1 = -5 + 0.25 * 25 = 1.25$$

Therefore $$i_2 = 0 + \frac{0.25(-5 + 1.25)}{2} = -0.46875$$

$$p_2 = f(t_1 + h, i_1 + hs_1, z_1 + hp_1)$$
$$= -5(z_1 + hp_1) - (i_1 + hs_1)$$
$$= -5(-5 + 0.25 * 25) - (0 + 0.25 * -5) = -5$$

$$z_2 = z_1 + \frac{h(p_1 + p_2)}{2} = -5 + \frac{0.25(25 + (-5))}{2}$$
$$= -2.5$$

At the end of this step, $i_2 = -0.46875$ and $z_2 = -2.5$

For the next step

$$s_2 = z_2 = -2.5$$

$$p_2 = -5z_2 - i_2 = -5(-2.5) - (-0.46875)$$
$$= 12.968875$$

$$s_3 = z_2 + hp_2 = -2.5 + 0.25(12.968875)$$
$$= 0.742219$$

$$i_3 = i_2 + \frac{h(s_2 + s_3)}{2} = -0.46875 + \frac{0.25(-2.5 + 0.742219)}{2}$$
$$= \textbf{-0.68847}$$

$$p_3 = f(t_2 + h, i_2 + hs_2, z_2 + hp_2)$$
$$= -5(z_2 + hp_2) - (i_2 + hs_2)$$
$$= -5(-2.5 + 0.25 * 12.968875)$$
$$- \left(-0.46875 + 0.25 * (-2.5)\right)$$
$$= -2.6173$$

$$z_3 = z_2 + \frac{h(p_2 + p_3)}{2}$$

$$= -2.5 + \frac{0.25\big(12.968875 + (-2.6173)\big)}{2}$$

$$= -1.2061$$

At the end of this step, $i_3 = -0.68847 \; and \; z_3 = -1.2061$

$Hence, i(0.5) = i_3 = -0.68847A$

CONSENT FOR PUBLICATION

Not applicable.

CONFLICT OF INTEREST

The author declares no conflict of interest, financial or otherwise.

ACKNOWLEDGEMENT

Declared none.

REFERENCES

[1]   V. Rajaraman, Computer Oriented Numerical Methods, Prentice-Hall of India, New Delhi, 1999.
[2]   S. S. Sastry, Introductory Methods of Numerical Analysis, Prentice-Hall of India, New Delhi, 2013.
[3]   E. Balagurusamy, Numerical Methods, Tata McGraw Hill Education, New Delhi, 2014.
[4]   S. C. Chapra & R. P. Canale, Numerical Methods for Engineers, McGraw Hill International Edition, Singapore, 2010.
[5]   D. G. Zill, A First Course in Differential Equations with Modeling Applications, Brooks/Cole, Cengage Learning, USA, 2012.
[6]   J. F. Epperson, An introduction to numerical methods and analysis, Second edition, John Wiley & Sons, Inc., Hoboken, New Jersey, 2013.

EXERCISES

(Q-01) A mathematical model for the area A (in cm^2) that the charge on one of the

plate of a capacitor is given by:

$$\frac{dA}{dt} = A(2.128 - 0.0432A)^*$$

Suppose that the initial area is $0.24 \ cm^2$.

(a) Use the Runge-Kutta Fourth Order method with $h = 0.5$ to complete the

following table:

t (days)	1	2	3	4	5
A (observed)	3.02	7.65	13.53	27.90	36.30
A (approximated)					

(b) Use a numerical solver to graph the solution of the initial-value problem. Estimate the values $A(1)$, $A(2)$, $A(3)$, $A(4)$, and $A(5)$ from the graph.

(Q-02) If air resistance is proportional to the square of the instantaneous velocity, then the velocity v of a mass m dropped from a given height is determined from

$$m\frac{dv}{dt} = mg - kv, k > 0$$

Let v(0) = 0 , k=0.125, m=5 slugs and g=32 ft/s^2.

Use all the methods for solving differential equations discussed in this chapter with $h = 1$ to approximate the velocity $v(5)$.

(Q-03) An R-L circuit was initially carrying 1mA of current when it is connected with voltage source of 10V (DC) at t=0. Calculate the current in the circuit at t=0.2 (R= 10 ohm and L = 2 H).

(Q-04)

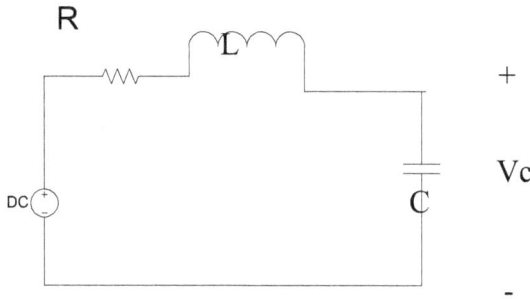

If the initial voltage of the capacitor is 5V and initial inductor current is 1A and $i_L = 0A$ then find the inductor current at t=2s.

(Q-05)

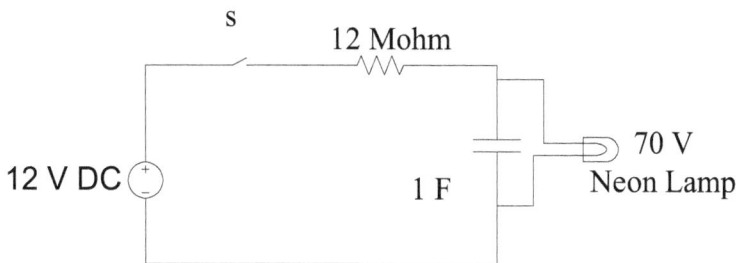

The above circuit is a delay circuit which is used to provide desired delay. A 70V neon lamp is connected across the capacitor. The lamp will glow when the voltage across the capacitor becomes greater than or equal to 70V. find the voltage across the capacitor at t= 2 ms. Assume the switch was open before t = 0s.

(Q-06) Consider a sample of a certain radioactive isotope. The atoms of such an isotope are unstable, with a certain proportion decaying each second. In particular, the mass M of the sample will decrease as atoms are lost, with the rate of decrease proportional to the number of atoms. We can write this as a differential equation

$$\frac{dM}{dt} = -rM$$

Where r is a constant of proportionality. With M(0) = 1 and r = 0.735, compute the mass of the sample at t = 0.02 using Euler's modified method.

(Q-07)

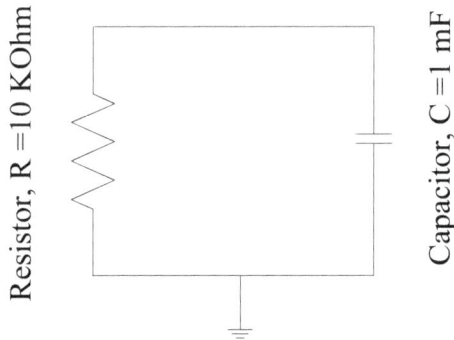

The above figure shows a simple kind of electric circuit known as an RC circuit.

It follows the differential equation as follows:

$$\left(-\frac{dQ}{dt}\right) R = \frac{Q}{C}$$

Where Q is the amount of charge held in the capacitor, R is the value of resistance and C is the value of capacitance. $Q_o = 0.5\,C$.

Find the Charge on the capacitor at t = 2 seconds.

(Q-08) The op amp integrator is used in numerous applications, especially in analog computers. An integrator is an op amp circuit whose output is proportional to the integral of the input signal.

The integrator using op-amp is shown as follows:

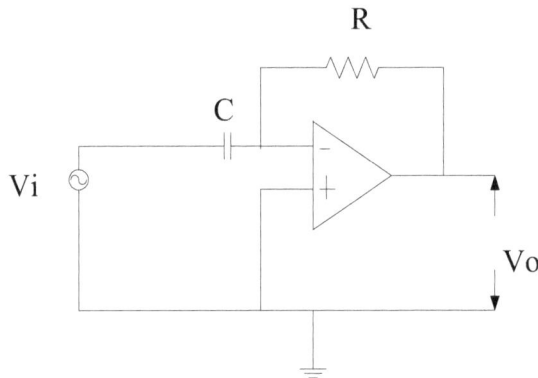

The above integrator circuit can be modeled mathematically by a first order differential equation. Derive the differential equation. Take,

$$v_i = 2t, \quad 0 < t < 5$$

And let $v_0(t = 0) = 2V$. Find the output voltage at t=2.5 seconds using all methods discussed in this chapter. Let C=0.5 F and L= 2 H

$$[\textit{Hints: } \frac{v_i}{R} = -C\frac{dv_o}{dt}]$$

(Q-09)

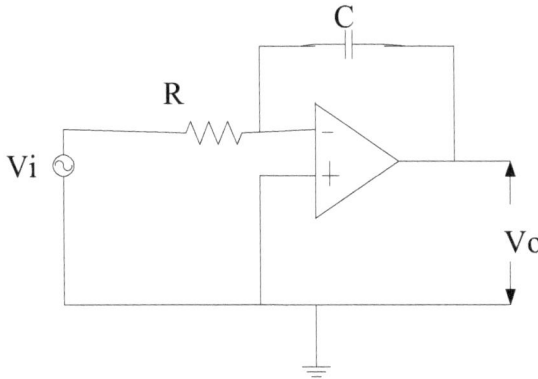

The above is a differentiator circuit which is also a first order circuit and can be mathematically modeled by a first order differential equation. Take the same values of the capacitor and resistor from problem (Q-09).

Derive the differential equation. Solve it analytically. And explain to other students, is the same concept of numerical solution is still applicable here or not. If it is applicable then discuss how? Feel free to choose any suitable initial value and any particular time.

(Q-10) Newton's law of cooling is a differential equation that predicts the cooling of a warm body placed in a cold environment. According to the law, the rate at which the temperature of the body decreases is proportional to the difference of temperature between the body and its environment. In symbols,

$$\frac{dT}{dt} = -k(T - T_e)$$

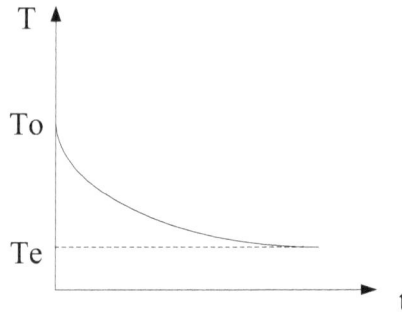

Cooling of a warm body (T $= T_e + Ce^{-kt}$)

where T is the temperature of the object, T_e is the (constant) temperature of the environment, and k is a constant of proportionality. Let T_e = 278K, k=5 . Find the temperature of the object at t =2 seconds. Take $T_o = 330\ K$. Use all the methods discussed in this chapter.

(Q-11) In chemistry, the rate at which a given chemical reaction occurs is often determined by a differential equation. For example, consider the decomposition of nitrogen dioxide:

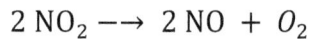

$$2\ NO_2 \dashrightarrow 2\ NO + O_2$$

Because this reaction requires two molecules of NO_2, the rate at which the reaction occurs is proportional to the square of the concentration of NO_2. That is,

$$\frac{d[NO_2]}{dt} = -k[NO_2]^2$$

where [NO2] is the concentration of NO_2 , and k is a constant. Solve the above differential equation using analytical methods of solving ordinary differential equations. Also calculate the concentration of NO_2 at t = 150 mS by considering that k =2 and initial concentration is 1000 moles/L.

(Q-12) Consider a colony of bacteria growing in an environment with limited resources. For example, there may be a scarcity of food, or space constraints on the size of the colony. In this case, it is not reasonable to expect the colony to grow exponentially—indeed, the colony will unable to grow larger than some maximum population P_{max}.

In this case, a common model for the growth of the colony is the logistic equation,

$$\frac{dP}{dt} = kP(1 - \frac{P}{P_{max}})$$

Let $P_{max} = 1000000$ and $k = 2.5$ and $P_0 = 35646$. Find the population at $t = 7$ minutes numerically using Euler's modified method and fourth order Runge-Kutta method.

(Q-13)

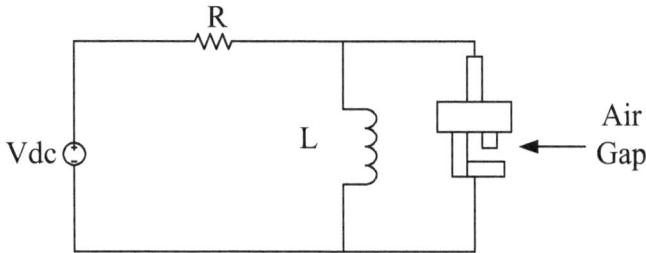

A solenoid with resistance 4 ohm and inductance 6 mH is used in an automobile ignition circuit similar to that in the above figure. Derive the first order differential equation for the ignition circuit. Solve it analytically. If the battery supplies 12 V, determine the current through the solenoid numerically using any suitable method at t = 0.9 second, assume initial current be 800 mA. Also compare the obtained result with the value obtained from the analytical solution.

[N.B: *Readers are advised to perform the above problems by using all the methods those have been discussed in this chapter.*]

CHAPTER 7

Introduction of Advanced Computational Methods

Md. Masud Rana[1*], **Wei Xu**[1] and **Youguang Guo**[1]

[1] *Department of Electrical & Electronic Engineering, Rajshahi University of Engineering & Technology, Bangladesh*

[2] *Department of Electrical & Electronic Engineering, Huazhong University of Science and Technology (HUST), China*

[3] *Department of Electrical & Electronic Engineering, University of Technology Sydney (UTS), Australia*

Abstract: In this chapter, advanced computational methods such as FD, FDTD, MoM, and FEM have been presented. Various types of computational methods are discussed with the engineering problem analysis.

Keywords: FD, FDTD, MoM, FEM, Algorithm, Engineering, Application.

INTRODUCTION

It is rare for real-life engineering problems, particularly electromagnetic problems, to fall neatly into a class that can be solved by the analytical methods presented in the preceding chapters. Classical approaches may fail if [1]: the partial differential equation (PDE) is not linear and cannot be linearized without seriously affecting the result, the solution region is complex, the boundary conditions are of mixed types, the boundary conditions are time-dependent, and the medium is inhomogeneous or anisotropic. Whenever a problem with such complexity arises, numerical solutions must be employed. Of the numerical methods available for solving PDEs are the finite difference (FD), finite difference time domain (FDTD), Method of Moment (MoM), and Finite Element Method (FEM), which are more easily understood, more frequently used, and more universally applicable than any other. These methods are discussed below.

---

*Corresponding author Md. Masud Rana:** Department of Electrical & Electronic Engineering, Rajshahi University of Engineering & Technology, Bangladesh; E-mail: md.masud.rana.ruet@gmail.com

Finite Difference (FD) Method

The finite difference (FD) method was first developed by A. Thom [2] in the 1920s under the title "the method of squares" to solve nonlinear hydrodynamic equations. Since then, the method has found applications in solving different field problems. The finite-difference techniques are based upon approximations that permit replacing differential equations with finite difference equations. These finite difference approximations are algebraic in form; they relate the value of the dependent variable at a point in the solution region to the values at some neighboring points. Thus, a finite-difference solution basically involves three steps:

(1) Dividing the solution region into a grid of nodes

(2) Approximating the differential equation and boundary conditions by a set of linear algebraic equations on the grid point in the solution region to its values at the neighboring points

(3) Solving the set of algebraic equations subjected to the prescribed boundary conditions and/or initial conditions.

Before finding the finite-difference solutions to specific PDEs, we will examine how one constructs finite difference approximations from a given differential equation. This essentially involves estimating derivatives numerically. The partial derivatives in the PDE at each grid point are approximated from neighboring values by using Taylor's theorem.

A more general approach is using Taylor's series. Let $f(x)$ has n continuous derivatives over the interval (a,b). Then for $a < x_0$, $x_0 + h < b$. According to the well-known expansion,

$$f(x_0 + \Delta x) = f(x_0) + \Delta x f'(x_0) + \frac{1}{2!}(\Delta x)^2 f''(x_0) + \frac{1}{3!}(\Delta x)^3 f'''(x_0) + \ldots\ldots + O(h^n) \quad \textbf{(7.1)}$$

and

$$f(x_0 - \Delta x) = f(x_0) - \Delta x f'(x_0) + \frac{1}{2!}(\Delta x)^2 f''(x_0) - \frac{1}{3!}(\Delta x)^3 f'''(x_0) + \ldots\ldots + O(h^n) \textbf{(7.2)}$$

The usual interpretation of Taylor's theorem says that if we know the value $f(x)$ and the values of its derivatives at a point x_0, then we can write down its value at a nearby point $x_0 + h$. Truncating (7.1) after the first derivatives term gives,

$$f(x_0 + \Delta x) \approx f(x_0) + \Delta x f'(x_0) \tag{7.3}$$

Rearranging (8.3) gives,

$$f'(x_0) \approx \frac{f(x_0 + \Delta x) - f(x_0)}{\Delta x} \tag{7.4}$$

This approximation is called forward FD approximation since we start at x_0 and step forwards to the point $x_0 + h$. h is called the step size $(h > 0)$. From the similar way of (7.2), we obtain:

$$f'(x_0) \approx \frac{f(x_0) - f(x_0 - \Delta x)}{\Delta x} \tag{7.5}$$

This approximation is called backward FD approximation. Upon adding these expansions (7.1) and (7.2), we obtain:

$$f(x_0 + \Delta x) + f(x_0 - \Delta x) = 2f(x_0) + (\Delta x)^2 f''(x_0) + \ldots\ldots + O(\Delta x)^4 \tag{7.6}$$

Where $O(\Delta x)^4$ is the error introduced by truncating the series. We say that this error is of the order $(\Delta x)^4$ or simply $O(\Delta x)^4$. Therefore, $O(\Delta x)^4$ represents terms that are not greater than $(\Delta x)^4$. Assuming that these terms are negligible,

$$f''(x_0) \approx \frac{f(x_0 + \Delta x) - 2f(x_0) + f(x_0 - \Delta x)}{(\Delta x)^2} \tag{7.7}$$

This approximation is called the central-difference FD approximation. We now construct common FD approximations to common partial derivatives. For simplicity, we suppose that φ is a function of only two variables t and x. As t is held constant, φ is effectively a function of the single variable x, so we can use Taylor's series formula (7.1), where the ordinary derivative terms are now partial derivatives and the argument (t, x) instead of x, so we can write the equation (7.1) as follows:

$$\varphi(t, x_0 + \Delta x) = \varphi(t, x_0) + \Delta x \varphi_x(t, x_0) + \frac{1}{2!}(\Delta x)^2 \varphi_{xx}(t, x_0) + \ldots\ldots + O(\Delta x^n) \quad \textbf{(7.8a)}$$

Truncating (7.8a) to $O(\Delta x)^2$ gives:

$$\varphi(t, x_0 + \Delta x) = \varphi(t, x_0) + \Delta x \varphi_x(t, x_0) + \ldots\ldots + O(\Delta x^2) \quad \textbf{(7.8b)}$$

Rearranging (7.8b) and neglecting the error terms gives:

$$\varphi_x(t, x_0) = \frac{\varphi(t, x_0 + \Delta x) - \varphi(t, x_0)}{\Delta x} \quad \textbf{(7.9)}$$

Equation (7.9) holds at any point (t, x_0). For solving PDEs, we are restricted to a grid of discrete x values $x_1, x_2, \ldots\ldots x_N$, and discrete t levels $0 = t_0, t_1, \ldots$. We will assume a constant grid spacing, Δx in x, so that $x_{i+1} = x_i + \Delta x$. Evaluating (7.9) for a point (t_n, x_i) on the grid gives:

$$\varphi_x(t_n, x_i) = \frac{\varphi(t_n, x_{i+1}) - \varphi(t_n, x_i)}{\Delta x} \quad \textbf{(7.10)}$$

We will use the common subscript/superscript notation.

$$\varphi_i^n = \varphi(t_n, x_i) \quad \textbf{(7.11)}$$

So (7.10) can be written as follows:

$$\varphi_x(t_n, x_i) \approx \frac{\varphi_{i+1}^n - \varphi_i^n}{\Delta x} \quad \textbf{(7.12)}$$

(7.12) is the first-order forward difference approximation to $\varphi_x(t_n, x_i)$. Similarly, the backward and central difference approximations can be written as follows:

$$\varphi_x(t_n, x_i) \approx \frac{\varphi_i^n - \varphi_{i-1}^n}{\Delta x} \quad \textbf{(7.13)}$$

and

$$\varphi_x(t_n, x_i) \approx \frac{\varphi_{i+1}^n - \varphi_{i-1}^n}{2\Delta x} \tag{7.14}$$

FD approximations to partial derivatives with respect to t are derived in a similar manner and are included in Table **7.1**.

Table 7.1. Finite difference Toolkit for partial derivatives.

Partial Derivatives	Finite difference Approximation	Order	Type
$\dfrac{\partial \varphi}{\partial x} = \varphi_x$	$\dfrac{\varphi_{i+1}^n - \varphi_i^n}{\Delta x}$	First in x	forward
$\dfrac{\partial \varphi}{\partial x} = \varphi_x$	$\dfrac{\varphi_i^n - \varphi_{i-1}^n}{\Delta x}$	First in x	backward
$\dfrac{\partial \varphi}{\partial x} = \varphi_x$	$\dfrac{\varphi_{i+1}^n - \varphi_{i-1}^n}{2\Delta x}$	Second in x	central
$\dfrac{\partial^2 \varphi}{\partial x^2} = \varphi_{xx}$	$\dfrac{\varphi_{i+1}^n - 2\varphi_i^n + \varphi_{i-1}^n}{\Delta x^2}$	Second in x	symmetric
$\dfrac{\partial \varphi}{\partial t} = \varphi_t$	$\dfrac{\varphi_i^{n+1} - \varphi_i^n}{\Delta t}$	First in t	forward
$\dfrac{\partial \varphi}{\partial t} = \varphi_t$	$\dfrac{\varphi_i^n - \varphi_i^{n-1}}{\Delta t}$	First in t	backward
$\dfrac{\partial \varphi}{\partial t} = \varphi_t$	$\dfrac{\varphi_i^{n+1} - \varphi_i^{n-1}}{2\Delta t}$	Second in t	central
$\dfrac{\partial^2 \varphi}{\partial t^2} = \varphi_{tt}$	$\dfrac{\varphi_i^{n+1} - 2\varphi_i^n + \varphi_i^{n-1}}{\Delta t^2}$	Second in t	symmetric

Example 7.1:   Solve the one-dimensional advection equation $\varphi_t + k\varphi_x = 0$, $p \leq x \leq q$ subject to $\varphi(0, x) = f(x)$ and use the finite difference method.

Solution:

The computational domain (Fig. **7.1**) contains an infinite number of x values, so first, we must replace them with a finite set. This process is called spatial discretisation.

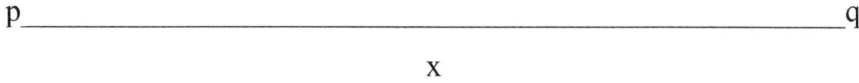

p_____q

$$x$$

Fig. (7.1). The computational domain contains an infinite number of x values.

For simplicity, the computational domain is replaced by a grid of N equally spaced grid points. Starting with the first grid point at $x = p$ and ending with the last grid point at $x = q$, the constant grid spacing Δx is

$$\Delta x = \frac{(q-p)}{(N-1)} \tag{7.15}$$

The values of x in the discretised computational domain are indexed by subscripts to give:

$$x_1 = p, x_2 = p + \Delta x, \ldots\ldots, x_i = p + (i-1)\Delta x, \ldots\ldots, x_N = p + (N-1)\Delta x = q,$$

Since the grid spacing is constant,

$$x_{i+1} = x_i + \Delta x \tag{7.16}$$

Fixing t at $t = t_n$, we approximate the spatial partial derivative φ_x, in (7.12a) at each point (t_n, x_i) using the forward difference formula from the toolkit in Table 7.1 to give,

$$\varphi_x \approx \frac{\varphi_{i+1}^n - \varphi_i^n}{\Delta x} \tag{7.17}$$

We obtain,

$$\varphi_t + k\frac{\varphi_{i+1}^n - \varphi_i^n}{\Delta x} = 0 \tag{7.18}$$

Eq. (7.18) is said to be in semi-discrete form since only the spatial derivative has been discretised. Now, fixing x at $x = x_i$, we approximate the temporal partial derivatives φ_t in (7.12a) at each point (t_n, x_i) using the forward difference formula from the toolkit in Table 7.1 (where Δt is the spacing between time levels) to give,

$$\frac{\varphi_i^{n+1} - \varphi_i^n}{\Delta t} \tag{7.19}$$

Eq. (7.18) can be written as

$$\frac{\varphi_i^{n+1} - \varphi_i^n}{\Delta t} + k \frac{\varphi_{i+1}^n - \varphi_i^n}{\Delta x} = 0 \tag{7.20}$$

Eq. (7.20) can be written as follows,

$$\varphi_i^{n+1} = \varphi_i^n - \frac{k \Delta t}{\Delta x} \left(\varphi_{i+1}^n - \varphi_i^n \right) \tag{7.21}$$

Exercise 7.2 Solve the one dimensional boundary value problem $-\varphi_{xx} = x^2$, $0 \le x \le 1$ subject to $\varphi(0) = 0 = \varphi(1)$. Use finite difference method.

Solution: Now we derive the FD equations of two most widely used elliptic PDEs are Poisson's and Laplace's equation. In general, the Poisson's is

$$\varphi_{xx} + \varphi_{yy} + \varphi_{zz} = -\frac{\rho_v}{\varepsilon} \tag{7.22}$$

For 2-D solution region such as in Fig. 7.2 (a), ρ_v is replaced by ρ_s, $\varphi_{zz} = 0$, so

$$\varphi_{xx} + \varphi_{yy} = -\frac{\rho_v}{\varepsilon} \tag{7.23}$$

Poisson's equation may be used to model a wide range of phenomena, including gravitational fields, stress patterns, simplified viscous flow, TEM transmission line modeling, and EM field. If the solution region is charge-free ($\rho_s = 0$), eq. (7.23) becomes Laplace's equation.

$$\varphi_{xx} + \varphi_{yy} = 0 \tag{7.24}$$

Laplace's equation may be used to model a wide range of applications, including steady-state groundwater flow, temperature distribution over a region, and potential flow, *etc.*

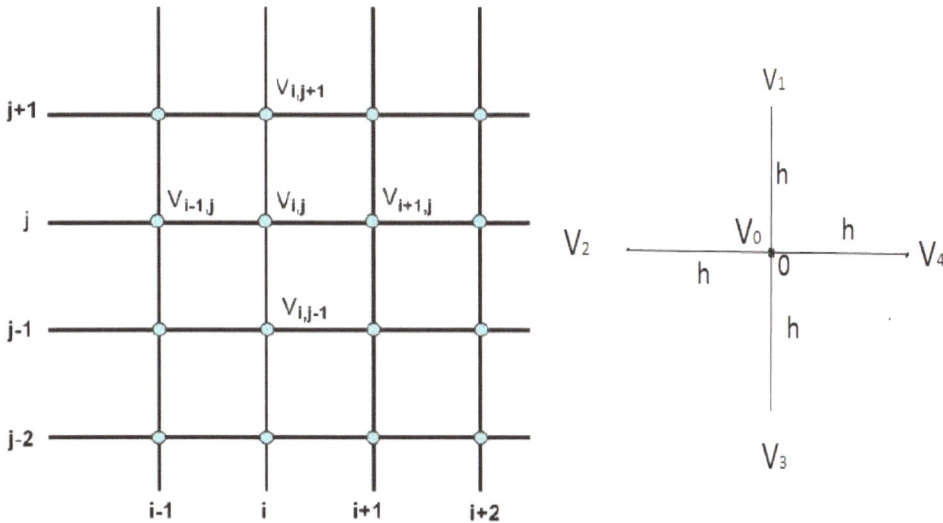

Fig. (7.2). Finite difference solution pattern: **(a)** computational grid, **(b)** finite-difference five node molecule.

The computational domain is discretised using a constant grid spacing of Δx and Δy in the x and y directions, respectively. Grid points are indexed by (i, j) in the usual way, and the approximate value of φ at grid point (i, j) is denoted by φ_{ij}. Fig. (**7.2a**) shows a rectangular grid with $M \times N$ grid points in the x and y directions, respectively. Each partial derivative in Fig. (**7.2b**) is replaced by a symmetrical FD approximation from our tool kit (Table. **7.1**) to give:

$$\frac{\varphi_{i+1,j} - 2\varphi_{i,j} + \varphi_{i-1,j}}{\Delta x^2} + \frac{\varphi_{i,j+1} - 2\varphi_{i,j} + \varphi_{i,j-1}}{\Delta y^2} = -\frac{\rho_s}{\varepsilon} \tag{7.25}$$

Letting $\Delta x = \Delta y = h$, From Eq. (8.25) we get

$$\varphi_{i+1,j} + \varphi_{i-1,j} + \varphi_{i,j+1} + \varphi_{i,j-1} - 4\varphi_{i,j} = -\frac{h^2 \rho_s}{\varepsilon} \tag{7.26}$$

$$\varphi_{i,j} = \frac{1}{4}(\varphi_{i+1,j} + \varphi_{i-1,j} + \varphi_{i,j+1} + \varphi_{i,j-1} + \frac{h^2 \rho_s}{\varepsilon}) \qquad (7.27)$$

Where, h is called mesh size. Equation (7.27) is the finite difference approximation to Poisson's equation. Similarly, for the charge-free region $\rho_s = 0$, equation (7.27) becomes Laplace's equation and can be written as follows:

$$\varphi_{i,j} = \frac{1}{4}(\varphi_{i+1,j} + \varphi_{i-1,j} + \varphi_{i,j+1} + \varphi_{i,j-1}) \qquad (7.28)$$

This equation is essentially a five-node finite difference approximation for the potential at the central point of a square mesh. Fig. **(7.2b)** illustrates the five node molecule. Thus, eq. (7.28) applied to the molecule becomes

$$\varphi_0 = \frac{1}{4}(\varphi_1 + \varphi_2 + \varphi_3 + \varphi_4) \qquad \textbf{(7.29)}$$

To apply eq. (7.27) or eq. (7.29) to a given problem, one of the following two methods is commonly used.

Direct Method/Band Matrix Method

Equation (7.29) applied to all free nodes results in a set of simultaneous equations of the form:

$$[A][\varphi] = [B] \qquad \textbf{(7.30)}$$

Where $[A]$ is a sparse matrix (one having many zero terms),   consists of the unknown potentials at the free nodes, and $[B]$ is another column matrix formed by the known potentials at the fixed nodes. Matrix $[A]$ is also banded in that its nonzero terms appear clustered near the main diagonal because neighboring nodes affect the potential at each node [3, 4]. The sparse, band matrix is easily inverted to determine $[\varphi]$. Thus we obtain the potential at the free nodes from the matrix $[\varphi]$ as

$$[\varphi] = [A]^{-1}[B] \qquad \textbf{(7.31)}$$

So, the standard Gaussian elimination method can be used as it is a direct method.

Iteration Method

For practical problems, $[A]$ is likely to be a large matrix, which makes the direct solution of (7.28) computationally inefficient. More efficient methods use iterative approaches where an initial estimate for is updated to form a better estimate. We start setting initial values at the free nodes equal to zero or any reasonable guessed value. Keeping the values at the fixed nodes unchanged at all times, we apply (7.28) to every free node in turn until the values at all free nodes are calculated. The process is repeated until a prescribed degree of accuracy is achieved or until the old and new values at each node are satisfactorily close. For Jacobi iteration, we introduce the iteration index as a superscript, m, and write (7.28) as the Jacobi formula

$$\varphi_{i,j}^{m+1} = \frac{1}{4}(\varphi_{i+1,j}^m + \varphi_{i-1,j}^m + \varphi_{i,j+1}^m + \varphi_{i,j-1}^m) \tag{7.32}$$

For each interior grid point (i, j), $\varphi_{i,j}$ at the next iteration $(m+1)$ is found from (7.32). Once iteration has been completed for all interior grid points, we compute φ^{m+1} and φ^m. If

$$\left|\varphi^{m+1} - \varphi^m\right| < \text{tol} \tag{7.33}$$

Where tol is a pre-defined tolerance, the iteration terminates, and the solution to (7.14) is φ^{m+1}; otherwise, the iteration continues.

Gauss-Seidel Iteration

This a potentially more efficient version of Jacobi iteration. We note in eq. (7.32), some of the updated $\varphi_{i,j}$ values are already available for use in the iteration formula.

So, according to the Gauss-Seidel method, eq. (7.32) can be written as follows:

$$\varphi_{i,j}^{m+1} = \frac{1}{4}(\varphi_{i-1,j}^{m+1} + \varphi_{i,j-1}^{m+1} + \varphi_{i+1,j}^m + \varphi_{i,j+1}^m) \tag{7.34}$$

The details about Jacobi and Gauss-Seidel methods with algorithms have been discussed in Chapter 3.

Example 7.2. Determine the potential at the free nodes in the potential system of Fig. (**7.3**) below using the FD method.

We first set the initial values of the potential at the free nodes equal to zero. We apply

$$V_0 = \frac{1}{4}(V_1 + V_2 + V_3 + V_4)$$

to each free node using the newest surrounding potentials each time the potential at that node is calculated.

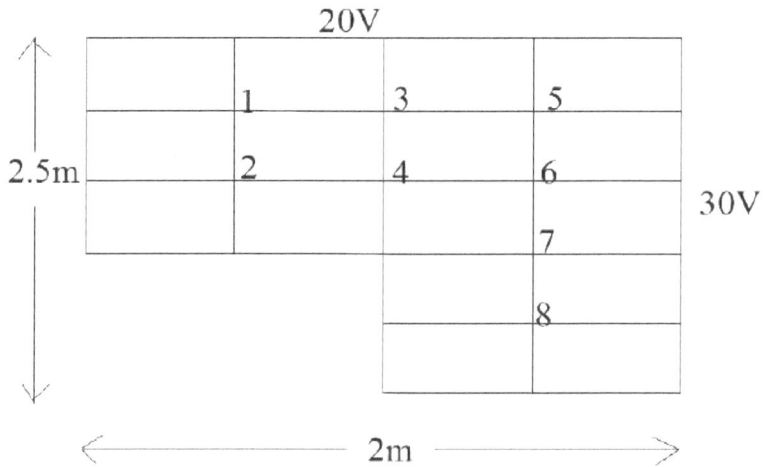

Fig. (7.2.1). Potential System.

For the first iteration:

$$V_1 = \frac{1}{4}(0 + 20 + 0 + 0) = 5$$

$$V_2 = \frac{1}{4}(5 + 0 + 0 + 0) = 1.25$$

$$V_3 = \frac{1}{4}(5 + 20 + 0 + 0) = 6.25$$

$$V_4 = \frac{1}{4}(1.25 + 6.25 + 0 + 0) = 1.875$$

and so on. To avoid confusion, each time a new value at a free node is calculated, we cross out the old value, as shown in Fig. **(7.2)**. After V8 is calculated, we start the second iteration at node 1:

$$V_1 = \frac{1}{4}(0 + 20 + 1.25 + 6.25) = 6.875$$

$$V_2 = \frac{1}{4}(6.875 + 0 + 0 + 1.875) = 2.187$$

and so on. If this process is continued, after 10 iterations, we obtain

$$V_1 = 10.04, \ V_2 = 4.956, V_3 = 15.22, V_4 = 9.786$$

$$V_5 = 21.05, \ V_6 = 18.97, V_7 = 15.06, V_8 = 11.25$$

Example 7.3 Use the iteration method to find the finite difference approximation to the potentials at points a and b of the following system (Fig. **7.3**).

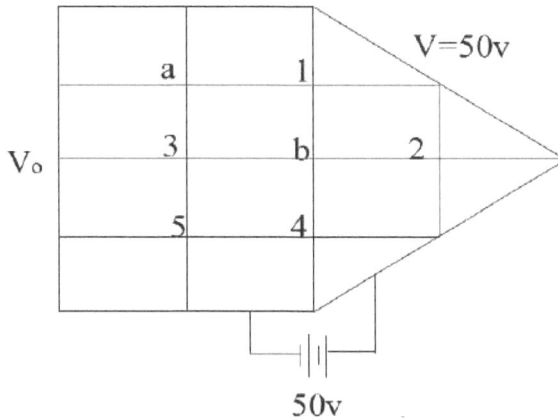

Fig. (7.3). Potential System.

We first set the initial values of the potential at the free nodes equal to zero. We apply

$$V_0 = \frac{1}{4}[V_1 + V_2 + V_3 + V_4]$$

to each free node using the newest surrounding potentials each time the potential at that node is calculated.

For the first iteration:

$$V_1 = \frac{1}{4}[50 + 0 + 0 + 50] = 25$$

$$V_2 = \frac{1}{4}[0 + 50 + 50 + 50] = 37.5$$

$$V_a = \frac{1}{4}[0 + 0 + 0 + 25] = 6.25$$

$$V_b = \frac{1}{4}[25 + 37.5 + 0 + 0] = 15.625$$

$$V_3 = \frac{1}{4}[0 + 6.25 + 0 + 15.625] = 5.4687$$

$$V_4 = \frac{1}{4}[0 + 50 + 50 + 15.625] = 28.9$$

$$V_5 = \frac{1}{4}[0 + 50 + 15.625 + 28.9] = 21.1$$

For the second iteration:

$$V_1 = \frac{1}{4}[50 + 50 + 6.25 + 15.625] = 30.47$$

$$V_2 = \frac{1}{4}[50 + 50 + 50 + 15.625] = 41.41$$

$$V_a = \frac{1}{4}[0 + 0 + 30.47 + 5.4687] = 8.985$$

$$V_b = \frac{1}{4}[30.47 + 41.41 + 5.4687 + 28.9] = 26.56$$

$$V_3 = \frac{1}{4}[0 + 8.985 + 26.56 + 21.1] = 14.162$$

$$V_4 = \frac{1}{4}[\,50 + 50 + 26.56 + 21.1\,] = 36.915$$

$$V_5 = \frac{1}{4}[\,0 + 50 + 14.162 + 36.915\,] = 25.27$$

Similarly, after 5 iteration, we obtain the desired result.

Finite Difference Time Domain (FDTD) Method

The starting point for the construction of the finite difference time domain (FDTD) method is Maxwell's curl equations. For a source-free region of space that is linear, isotropic, and non-dispersive, the differential form of Maxwell's equations which includes magnetic and electrical conductivity, are given by:

$$\frac{\partial \vec{E}}{\partial t} = \frac{1}{\varepsilon} \nabla \times \vec{H} - \frac{\sigma}{\varepsilon} \vec{E} \tag{7.35a}$$

$$\frac{\partial \vec{H}}{\partial t} = -\frac{1}{\mu} \nabla \times \vec{E} - \frac{\sigma_m}{\varepsilon} \vec{H} \tag{7.35b}$$

Where, $\vec{E}$ is the electric field strength vector in volts per meter, $\vec{H}$ is the magnetic field strength vector in amperes per meter and σ, σ_m are electrical and magnetic conductivity respectively. Equation (7.35) consists of two vector equations, and each vector equation can be composed into three scalar equations for three-dimensional spaces. Therefore, these vector curl equations (7.35) can be expanded into six-coupled scalar equations in a Cartesian coordinate system (x, y, z) as follows:

$$\frac{\partial E_x}{\partial t} = \frac{1}{\varepsilon}\left(\frac{\partial H_z}{dy} - \frac{\partial H_y}{dz} - \sigma E_x \right) \tag{7.36a}$$

$$\frac{\partial E_y}{\partial t} = \frac{1}{\varepsilon}\left(\frac{\partial H_x}{dz} - \frac{\partial H_z}{dx} - \sigma E_y \right) \tag{7.36b}$$

$$\frac{\partial E_z}{\partial t} = \frac{1}{\varepsilon}\left(\frac{\partial H_y}{dx} - \frac{\partial H_x}{dy} - \sigma E_z\right) \qquad (7.36c)$$

$$\frac{\partial H_x}{\partial t} = \frac{1}{\mu}\left(\frac{\partial E_y}{dz} - \frac{\partial E_z}{dy} - \sigma_m H_x\right) \qquad (7.36d)$$

$$\frac{\partial H_y}{\partial t} = \frac{1}{\mu}\left(\frac{\partial E_z}{dx} - \frac{\partial E_x}{dz} - \sigma_m H_y\right) \qquad (7.36e)$$

$$\frac{\partial H_z}{\partial t} = \frac{1}{\mu}\left(\frac{\partial E_x}{dy} - \frac{\partial E_y}{dx} - \sigma_m H_z\right) \qquad (7.36f)$$

Yee's notation for a space point within a rectangular lattice will also be used here and is as follows

$$(i, j, k) = (i\Delta x, j\Delta y, k\Delta z) \qquad (7.37)$$

and a function of space and time evaluated at a discrete point, and time is denoted as

$$U\mid_{i,j,k}^{n} = U(i\Delta x, j\Delta y, k\Delta z, n\Delta t) \qquad (7.38)$$

where Δx, Δy, and Δz are the spatial increment steps in the x, y, z dimensions respectively and i, j, and k are integers. Δt is the temporal increment and n is an integer.

According to the Yee algorithm [5], the space and time derivatives in Maxwell's equations can be replaced with the central finite difference expressions, which are second-order accurate in space and time. His algorithm interleaves the components of $\vec{E}$ and $\vec{H}$ in the space lattice at intervals of $\Delta x / 2$, $\Delta y / 2$ and $\Delta z / 2$ in the x, y, and z directions, respectively, as shown in the Yee cube of Fig. 7.3. Using this arrangement, the field components required to update Maxwell's equations is positioned adjacent to each other, viz., every $\vec{E}$ component is surrounded by four circulating $\vec{H}$ components, and every $\vec{H}$ component is surrounded by four circulating $\vec{E}$. By using the second-order accurate central finite difference approximation, the spatial partial derivative of U in the x-direction can be expressed as follows:

$$\frac{\partial(U\mid_{i,j,k}^{n})}{\partial x} = \frac{U\mid_{i+1/2,j,k}^{n} - U\mid_{i-1/2,j,k}^{n}}{\Delta x} + o(\Delta x^2) \tag{7.39}$$

The $\pm 1/2$ increment on the i lattice position designates a space finite difference over $\pm 1/2\Delta x$. The $\vec{E}$ and $\vec{H}$ components are located in discrete time in what is called a leapfrog arrangement. Yee interleaved the $\vec{E}$ and $\vec{H}$ fields at time intervals of $1/2\Delta t$ in order to implement his algorithm. This arrangement allows the $\vec{E}$ fields to be calculated and stored in memory, then at a time interval $1/2\Delta t$ seconds later, the $\vec{H}$ fields are calculated from the recently updated $\vec{E}$ fields and stored. The cycle can then continue until the total observation time reaches its end. To enable this arrangement of leapfrog time-stepping Yee defined the temporal partial derivative of U, at a fixed lattice point (i, j, k) as:

$$\frac{\partial(U\mid_{i,j,k}^{n})}{\partial t} = \frac{U\mid_{i,j,k}^{n+1/2} - U\mid_{i,j,k}^{n-1/2}}{\Delta t} + o(\Delta t^2) \tag{7.40}$$

Yee's original formulation assumed a lossless medium. For this analysis, the magnetic loss and electrical conductivity will not be neglected, requiring an approximation for the field component associated with the loss terms of (7.35). This term is estimated by the so-called semi-implicit approximation [2,3].

$$U\mid_{i,j,k}^{n} = \frac{U\mid_{i,j,k}^{n+1/2} + U\mid_{i,j,k}^{n-1/2}}{2} \tag{7.41}$$

The spatial and temporal partial derivatives and semi-implicit approximations are then substituted in Maxwell's curl equations of (7.36a)-(7.36f).

These resulting equations are then rearranged to solve for the current field value of the temporal derivative, which is $(n+1/2)\Delta t$ for the $\vec{H}$ components and $(n+1)\Delta t$ for the $\vec{E}$ components. Due to the availability of the FDTD updating equation in the literature, the updating equations for the FDTD method are not provided here. To get the updating equations of the FDTD method, the author recommends to see ref [5]. Position of electric and magnetic field components in Yee's grid is shown in Fig. (**7.4**).

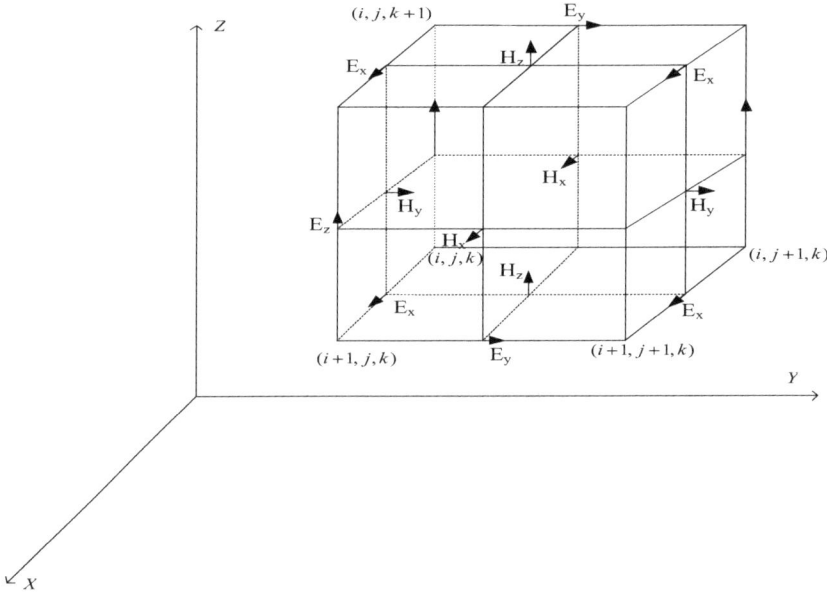

Fig. (7.4). Position of electric and magnetic field components in Yee's grid.

Update Equation

The explicit finite difference approximation of magnetic fields of (7.36) can be obtained as:

$$H_x\,\big|_{i,j+1/2,k+1/2}^{n+1/2} = H_x\,\big|_{i,j+1/2,k+1/2}^{n-1/2} + \frac{\Delta t}{\mu}\left[\frac{E_y\,\big|_{i,j+1/2,k+1}^{n} - E_y\,\big|_{i,j+1/2,k}^{n}}{\Delta z}\right]$$

$$-\frac{\Delta t}{\mu}\left[\frac{E_z\,\big|_{i+1/2,j,k+1}^{n} - E_z\,\big|_{i+1/2,j,k}^{n}}{\Delta y}\right]$$

$$(7.42a)$$

$$H_y\Big|_{i+1/2,j,k+1/2}^{n+1/2} = H_y\Big|_{i+1/2,j,k+1/2}^{n-1/2} + \frac{\Delta t}{\mu}\left[\frac{E_z\Big|_{i+1,j,k+1/2}^{n} - E_z\Big|_{i,j,k+1/2}^{n}}{\Delta x}\right]$$

$$-\frac{\Delta t}{\mu}\left[\frac{E_x\Big|_{i+1/2,j,k-1}^{n} - E_x\Big|_{i+1/2,j,k}^{n}}{\Delta z}\right] \tag{7.42b}$$

$$H_z\Big|_{i+1/2,j+1/2,k}^{n+1/2} = H_z\Big|_{i+1/2,j+1/2,k}^{n-1/2} + \frac{\Delta t}{\mu}\left[\frac{E_x\Big|_{i+1/2,j-1,k}^{n} - E_x\Big|_{i+1/2,j,k}^{n}}{\Delta y}\right]$$

$$-\frac{\Delta t}{\mu}\left[\frac{E_y\Big|_{i+1,j+1/2,k}^{n} - E_y\Big|_{i,j+1/2,k}^{n}}{\Delta x}\right] \tag{7.42c}$$

Similar equations for an electric field will get from Maxwell's equations. The update procedure of the FDTD method is summarized in Fig.(7.5).

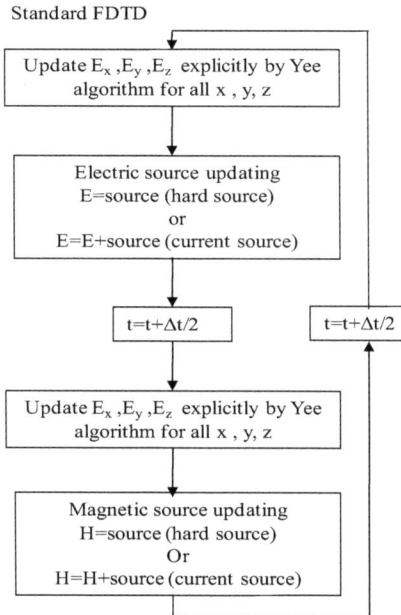

Standard FDTD

Update E_x, E_y, E_z explicitly by Yee algorithm for all x, y, z

Electric source updating
E=source (hard source)
or
E=E+source (current source)

t=t+Δt/2   t=t+Δt/2

Update E_x, E_y, E_z explicitly by Yee algorithm for all x, y, z

Magnetic source updating
H=source (hard source)
Or
H=H+source (current source)

Fig. (7.5). Flowcharts of the time-stepping process for standard FDTD method.

Selections of Grid Size and Time Step Size

In spite of the fact that the FDTD is popular, it is very memory and CPU-time intensive. Such intensive memory and CPU time requirements are due to the two physical constraints.

1) Spatial increment steps must be small enough in comparison with the wavelength (usually 10-20 steps per wavelength)

2) Time step must be small enough. More specifically, the CFL stipulates that [5]

$$u_{max} \Delta t \leq \left(\frac{1}{\Delta x^2} + \frac{1}{\Delta y^2} + \frac{1}{\Delta z^2} \right)^{-1/2} \qquad (7.43)$$

Absorbing Boundary Conditions

A basic consideration with the FDTD method to solving electromagnetic wave interaction problems is that many geometries of interest are defined in open regions where the spatial domain of the computed field is unbounded in one or more coordinate directions. More clearly, no computer can store an unlimited amount of data, and therefore, the field computation domain must be limited in size. The computational domain must be large enough to enclose the structure of interest, and a suitable boundary condition on the outer perimeter of the domain must be used to simulate its extension to infinity. However, to model open region problems, a boundary condition that would produce negligible reflections is necessary to truncate the computational domain. Schematic of a typical wave-equation problem and the same problem, where space has been truncated to some computational region are shown in Fig. (**7.6a**) and (**7.6b**) respectively.

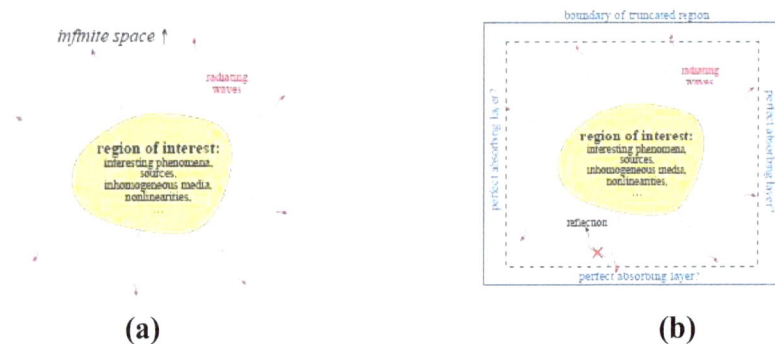

(a)   (b)

Fig. (7.6). (a) Schematic of a typical wave-equation problem, **(b)** The same problem, where space has been truncated to some computational region.

The various type ABCs which used for absorbing outward propagating wave are as below.

- One-way wave equation (Mur's ABC)
- Berenger Perfectly Matched Layer
- Uniaxial Perfectly Matched Layers (UPML)
- Convolutional Perfectly Matched Layer (CPML)

Limitations of FDTD Method

Since the FDTD method requires that the entire computational domain be gridded, and the grid spatial discretization must be sufficiently fine to resolve both the smallest electromagnetic wavelength and the smallest geometrical feature in the model, very large computational domains can be developed, which results in very long solution times. Models with long, thin features (like wires) are difficult to model in FDTD because of the excessively large computational domain required. There is no way to determine unique values for permittivity and permeability at a material interface. Space and time steps must satisfy the CFL condition, or the leapfrog integration used to solve the partial differential equation is probable to become unstable since the FDTD finds the E/H fields directly everywhere in the computational domain. If the field values at some distance are desired, it is likely that this distance will force the computational domain to be excessively large. Far-field extensions are available for FDTD but require some amount of post-processing.

Method of Moment (MoM)

Method of Moments transforms the integrodifferential equations into matrix systems of linear equations, which can be solved using digital computers. Let us consider the inhomogeneous equation:

$$L(u) = g \qquad\qquad (7.44a)$$

$$\Rightarrow L(u) - g = 0 \qquad\qquad (7.44b)$$

where, L is an operator which may be differential, integral or integrodifferential, g is the known function or excitation, and u is the unknown function to be determined. For example, consider the integral equation for a line charge density,

$$V_0 = \int \frac{\rho(x')dx'}{4\pi\varepsilon_0 r(x,x')}, \quad \text{then} \quad L = \int \frac{dx'}{4\pi\varepsilon_0 r(x,x')}, \quad g = V_0, \text{ and } u = \rho(x') \text{ . Similarly,}$$

consider the differential equation of the form:

$$-\frac{d^2 f(x)}{dx^2} = 3 + 3x^3 \text{ , where } g = 3 + 3x^3, \ L = -\frac{d^2}{dx^2} \text{ , and } u = f(x). \text{ So, to solve } u$$

using MoM, approximate it by the sum of weighted known basis functions or expansion functions, *i.e.*

$$u \cong \sum_{n=1}^{N} u_n = \sum_{n=1}^{N} I_n b_n, \qquad \text{where,} n = 1,2,......,N \qquad (7.45)$$

where, b_n is the expansion or basis function, I_n is its unknown complex coefficients to be determined, and N should be infinite theoretically and but practically it should be a finite number. The basis functions are chosen to model the expected behavior of the unknown function throughout its domain and can be scalars or vectors depending on the problem. If the basis functions have local support in the domain, they are called local or subsectional basis functions. If the support spans the entire problem domain, they are called global or entire domain basis function. Since L is linear, substitution of the equ. (7.45) into the (7.44a), we obtain,

$$L\left(\sum_{n=1}^{N} I_n b_n\right) \approx g \qquad (7.46)$$

where the error or residual is given by

$$R = g - L\left(\sum_{n=1}^{N} I_n b_n\right) \qquad (7.47)$$

Let us now generalize the method by which the boundary conditions are enforced. We define a set of weighting function or testing functions $w_1, w_2,......w_N$ in the range of L and take an inner product or moment between a basis function b_n and a testing or weighting function w_m. Requiring the inner product of each testing function with the residual function to be zero yields:

$$\langle w_m, (L(u_n) - g) \rangle = 0 \quad \text{where } m = 1,2,......,M \qquad (7.48a)$$

$$\left\langle w_m, L(\sum_{n=1}^{N} I_n b_n) \right\rangle = \left\langle w_m, g \right\rangle \quad \text{where } m = 1,2,......,M \tag{7.48b}$$

Where M should be infinite theoretically and but practically it should be a finite number. Since I_n is a constant, we can take it outside the inner product and equ. (7.48b) can be written as follows:

$$\sum_{n=1}^{N} I_n \left\langle w_m, L(b_n) \right\rangle = \left\langle w_m, g \right\rangle \quad \text{where } m = 1,2,......,M \tag{7.49}$$

The system can be written in matrix form as:

$$[A_{mn}][I_n] = [g_m] \tag{7.50}$$

where

$$[A_{mn}] = \begin{pmatrix} \langle w_1, L(b_1) \rangle & \langle w_1, L(b_2) \rangle & ...... \\ \langle w_2, L(b_1) \rangle & \langle w_2, L(b_2) \rangle & ...... \\ \vdots & \vdots & \ddots \end{pmatrix}, \quad [I_n] = \begin{pmatrix} I_1 \\ I_2 \\ \vdots \end{pmatrix} \text{ and } [g_m] = \begin{pmatrix} \langle w_1, g \rangle \\ \langle w_2, g \rangle \\ \vdots \end{pmatrix}$$

If the matrix $[A_{mn}]$ is not singular, the unknowns $[I_n]$ are simply obtained by

$$[I_n] = [A_{mn}]^{-1}[g_m] \tag{7.51}$$

and the original function u can be reconstructed using Eq. (7.45). The basis functions used previously can be defined as (i) pulse function, (ii) piecewise triangular function, (iii) piecewise sinusoidal function, and (iv) entire domain function. In this book, we use mostly pulse functions.

$$\text{The pulse basis functions}: b_n = \begin{cases} 1 & \text{if } x \text{ belongs to the interval } n \\ 0 & \text{otherwise} \end{cases}$$

For the testing or weighting functions, there are two methods such as (i) point matching or point collocation and (ii) Galerkin's method are used to enforce boundary conditions in the integral equation. In the point matching method, a delta function is used as a weighting function such $w_m(r) = \delta(r)$. On the other hand, in Galerkin's method, we are free to use whatever functions we wish for testing. For many problems, the choice of testing function is crucial to obtaining a good solution. In Galerkin's method, the basis functions themselves are used as the testing functions.

Example 7.4 Determine the charge density ρ_l along a thin conducting wire (radius a, length $L(L \gg a)$) as shown below using MoM.

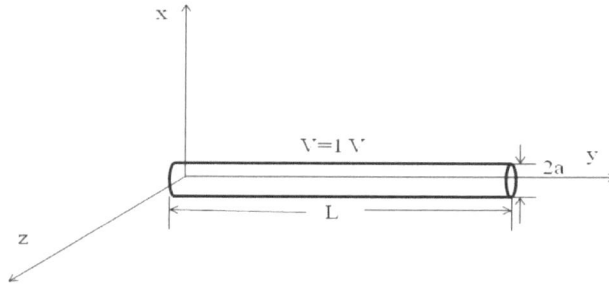

Fig. (7.7). Thin *conducting wire (radius a, length $L(L \gg a)$)*

Consider a thin conducting wire of radius a, length $L(L \gg a)$ located in free space along the y-axis as shown in Fig. **(7.7)**. Let the wire be maintained at a potential of $V_0 = 1V$. Our goal is to determine the charge density ρ_l along the wire using the moment method.

At any point on the wire, the integral solution of Poisson's equation

$$V_0 = \frac{1}{4\pi\varepsilon_0} \int_0^l \frac{\rho_l(y')dy'}{R(y,y')} \tag{7.48}$$

where $R(y,y') = R(\vec{r},\vec{r}')\big|_{x=z=0} = \sqrt{(y-y')^2 + (x)^2 + (z)^2} = \sqrt{(y-y')^2 + (a)^2}$

It is necessary to solve the integral equation to find the unknown function $\rho_l(y')$.

The solution may be obtained numerically by reducing the integral equation into a series of linear algebraic equations that may be solved by conventional matrix techniques.

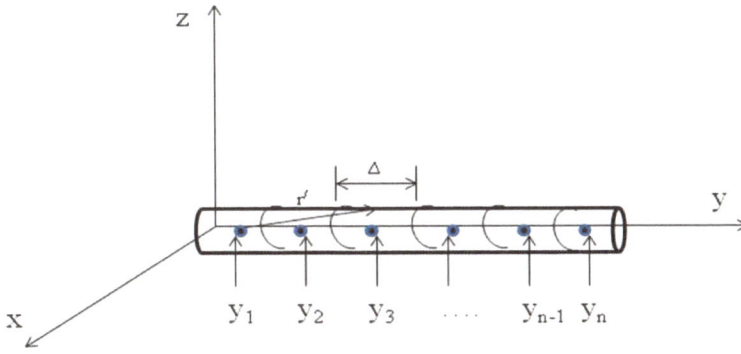

(a)

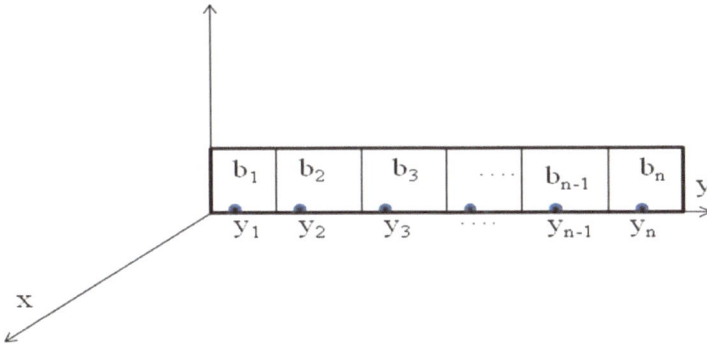

(b)

Fig. (7.8). (a) segmentation: $y_1, y_2, \ldots, y_N$ are observation points and r' shows a source point and **(b)** division of the charged strip into N sections.

Approximating the unknown charge density $\rho_l(y')$ by an expansion of N known basis functions with unknown coefficients as

$$\rho_l(y') = \sum_{n=1}^{N} I_n b_n, \qquad \text{where}, n = 1, 2, \ldots, N \qquad (7.49)$$

Integral equation after substituting this is

$$4\pi\varepsilon_0 = \int_0^l \frac{\sum_{n=1}^{N} I_n b_n(y')dy'}{R(y,y')} = \sum_{n=1}^{N} I_n \int_0^l \frac{b_n(y')dy'}{R(y,y')} \tag{7.50}$$

Now we have divided the wire into N uniform segments, each of length Δ as shown in Fig.(**7.8a**). We will choose our basis function as pulse function as follows:

$$b_n = \begin{cases} 1 & \text{for } (n-1)\Delta \le y \le n\Delta \\ 0 & \text{otherwise} \end{cases} \tag{7.51}$$

Now, let us apply the testing functions as delta function $[\delta(y-y_m)]$ for point matching and integration of any function with this delta function will give us the function value at $y = y_m$. Replacing observation variable y by a fixed point such as y_m results in an integrand that is solely a function of y'. It leads to an equation with N unknowns as follows:

$$4\pi\varepsilon_0 = I_1 \int_0^\Delta \frac{b_1(y')dy'}{R(y_m,y')} + I_2 \int_\Delta^{2\Delta} \frac{b_2(y')dy'}{R(y_m,y')} + \dots\dots$$
$$\dots\dots + I_n \int_{(n-1)\Delta}^{n\Delta} \frac{b_n(y')dy'}{R(y_m,y')} + \dots + I_N \int_{(N-1)\Delta}^{l} \frac{b_N(y')dy'}{R(y_m,y')} \tag{7.52}$$

Solution for these N unknowns constant, N linearly independent equations are required. N equations may be produced by choosing an observation point y_m on the wire where $m = 1,2,3,\dots\dots,N$ and at the center of each Δ length element, as shown in Fig. (**7.8b**). Result in equations of the form of the previous equation corresponding to each observation point as follows:

$$4\pi\varepsilon_0 = I_1 \int_0^\Delta \frac{b_1(y')dy'}{R(y_1,y')} + I_2 \int_\Delta^{2\Delta} \frac{b_2(y')dy'}{R(y_1,y')} + \dots\dots\dots\dots + I_N \int_{(N-1)\Delta}^{l} \frac{b_N(y')dy'}{R(y_1,y')}$$

$$4\pi\varepsilon_0 = I_1 \int_0^\Delta \frac{b_1(y')dy'}{R(y_2,y')} + I_2 \int_\Delta^{2\Delta} \frac{b_2(y')dy'}{R(y_2,y')} + \dots\dots\dots\dots + I_N \int_{(N-1)\Delta}^{l} \frac{b_N(y')dy'}{R(y_2,y')}$$

$$\vdots$$

$$4\pi\varepsilon_0 = I_1 \int_0^{\Delta} \frac{b_1(y')dy'}{R(y_N, y')} + I_2 \int_{\Delta}^{2\Delta} \frac{b_2(y')dy'}{R(y_N, y')} + \ldots\ldots + I_N \int_{(N-1)\Delta}^{l} \frac{b_N(y')dy'}{R(y_N, y')} \quad \textbf{(7.53)}$$

We may write the above equations in matrix form as

$$\begin{pmatrix} Z_{11} & Z_{12} & \ldots\ldots Z_{1N} \\ Z_{21} & Z_{22} & \ldots\ldots Z_{2N} \\ \vdots & & \vdots \\ Z_{N1} & Z_{N2} & \ldots\ldots Z_{NN} \end{pmatrix} \begin{pmatrix} I_1 \\ I_2 \\ \vdots \\ I_N \end{pmatrix} = \begin{pmatrix} V_0 \\ V_0 \\ \vdots \\ V_0 \end{pmatrix} \cong [Z_{mn}][I_n] = [V_m] \quad \textbf{(7.54)}$$

where $[v_m] = [4\pi\varepsilon_0]$

$$Z_{mn} = \int_0^l \frac{b_n(y')dy'}{\sqrt{(y_m - y')^2 + a^2}} = \int_{y_{n-1}}^{y_n} \frac{dy'}{\sqrt{(y_m - y')^2 + a^2}}$$

$$\cong \int_{y_{n-1}}^{y_n} \frac{dy'}{\sqrt{(y_m - y')^2}} = \int_{y_{n-1}}^{y_n} \frac{dy'}{y_m - y'} \approx \frac{\Delta}{|y_m - y_n|} \quad \text{for } m \neq n \quad \textbf{(7.55)}$$

Special care is taken for calculating the Z_{mn} for $m = n$, because the expression Z_{mn} is infinite for this case. Extraction of this singularity by substituting $y_m - y' = \xi \Rightarrow d\xi = -dy'$

$$Z_{mm} = -\int_{\Delta}^{0} \frac{d\xi}{\sqrt{(\xi)^2 + a^2}} = \int_0^{\Delta} \frac{d\xi}{\sqrt{(\xi)^2 + a^2}}$$

$$= \log\left(\xi + \sqrt{(\xi)^2 + a^2}\right) = \ln\left[\frac{\Delta + \sqrt{(\Delta)^2 + a^2}}{a}\right] \quad \textbf{(7.56)}$$

If $\Delta \gg a$, Then we get $Z_{mn} = \ln\dfrac{2\Delta}{a} \cong 2\ln\dfrac{\Delta}{a}$. So, the matrix form equation is

$$
\begin{bmatrix}
2\ln\left(\frac{\Delta}{a}\right) & \frac{\Delta}{|y_1-y_2|} & \cdots & \frac{\Delta}{|y_1-y_N|} \\
\frac{\Delta}{|y_2-y_1|} & 2\ln\left(\frac{\Delta}{a}\right) & \cdots & \frac{\Delta}{|y_2-y_N|} \\
\vdots & \vdots & & \vdots \\
\frac{\Delta}{|y_N-y_1|} & \frac{\Delta}{|y_N-y_2|} & \cdots & \frac{\Delta}{|y_N-y_N|}
\end{bmatrix}
\begin{bmatrix}
\rho_1 \\ \rho_2 \\ \vdots \\ \rho_N
\end{bmatrix}
=
\begin{bmatrix}
1 \\ 1 \\ \vdots \\ 1
\end{bmatrix}
\times 4\pi\varepsilon_0 V_0 \qquad (7.57)
$$

Using Vo = 1 V, L = 1 m, a = 1 mm, and $N = 10$ ($\Delta = \frac{L}{N}$), a Matlab code such as in Fig. **(7.9)** can be developed. The program in Figure is self-explanatory. It inverts matrix [A] and plots P_L against y.

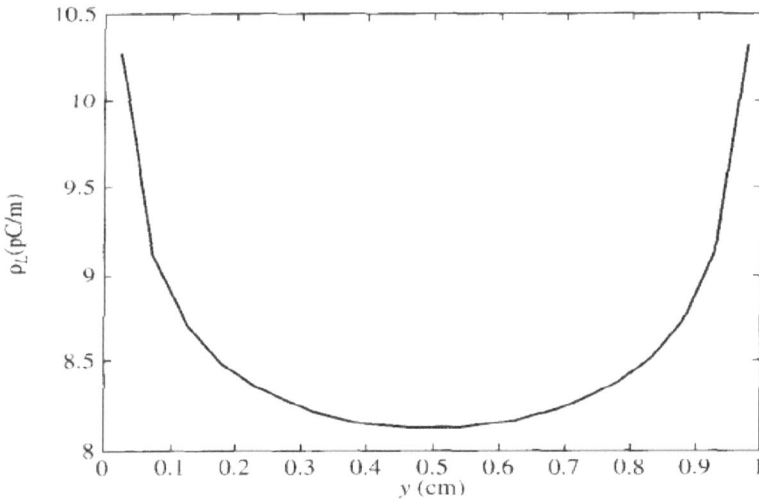

Fig. (7.9). Surface charge density over the wire surface.

Again, the total charge on the wire is

$$Q = \int \rho_L \, dl \qquad (7.58)$$

which can be written in discrete form as

$$Q = \sum_{k=1}^{N} \rho_k \Delta \qquad (7.59)$$

Using MATLAB Code, we get from figure Q = 8.536 pc.

Example 7.5 Consider a 1-D differential equation

$$-\frac{d^2 f(x)}{dx^2} = 3 + 2x^2$$

subject to the boundary condition $f(0) = f(1) = 0$. Solve the differential equation using Galerkin's MoM. $L(u) = g$

Solution: For this case, $u = f(x)$, $g = 3 + 2x^2$ and $L = -\frac{d^2}{dx^2}$. According to the nature of the known function, it is natural to choose the basis function as $b_n(x) = x^n$. However, the boundary condition $f(1) = 0$ can't be satisfied with such a basis function. A suitable basis function for this differential equation taking into account this boundary condition, is

$$b_n(x) = x - x^{n+1}; n = 1,2,......N$$

Assume $N = 2$, the total number of the subsection on the interval is $[0,1]$. Approximation of the unknown function is

$$f(x) \cong I_1 b_1(x) + I_2 b_2(x) = I_1(x - x^2) + I_2(x - x^3)$$

For Galerkin's MoM, the weighting functions are:

$$w_m(x) = x - x^{m+1}; m = 1,2,......M$$

Choosing a square $[Z]$ matrix where $M = N = 2$

$$Z_{11} = \langle w_1, L(b_1) \rangle = \int_0^1 w_1(x)L(b_1(x))dx = \int_0^1 (x - x^2)(2)dx = \frac{1}{3}$$

$$Z_{12} = \langle w_1, L(b_2) \rangle = \int_0^1 w_1(x)L(b_2(x))dx = \int_0^1 (x - x^2)(6x)dx = \frac{1}{2}$$

$$Z_{21} = \langle w_2, L(b_1) \rangle = \int_0^1 w_2(x)L(b_1(x))dx = \int_0^1 (x - x^3)(2)dx = \frac{1}{2}$$

$$Z_{22} = \langle w_2, L(b_2) \rangle = \int_0^1 w_2(x)L(b_2(x))dx = \int_0^1 (x - x^3)(6x)dx = \frac{4}{5}$$

for the calculation of V

$$V_1 = \langle k, w_1 \rangle = \int_0^1 k(x)w_1(x)dx = \int_0^1 (3 + 2x^2)(x - x^2)dx = \frac{3}{5}$$

$$V_2 = \langle k, w_2 \rangle = \int_0^1 k(x)w_2(x)dx = \int_0^1 (3 + 2x^2)(x - x^3)dx = \frac{11}{12}$$

Therefore,

$$[Z][I] = [V] \Rightarrow \begin{bmatrix} \frac{1}{3} & \frac{1}{2} \\ \frac{1}{2} & \frac{4}{5} \end{bmatrix} \begin{bmatrix} I_1 \\ I_2 \end{bmatrix} = \begin{bmatrix} \frac{3}{5} \\ \frac{11}{12} \end{bmatrix}$$

$$\Rightarrow [I] = \begin{bmatrix} I_1 \\ I_2 \end{bmatrix} = \begin{bmatrix} \frac{13}{10} \\ \frac{1}{3} \end{bmatrix}$$

The unknown function $f(x)$

$$f(x) \cong I_1(x - x^2) + I_2(x - x^3) = \frac{13}{10}(x - x^2) + \frac{1}{3}(x - x^3)$$

Finite Element Method (FEM)

The finite element method (FEM) is a numerical technique for solving problems that are described by partial differential equations or can be formulated as functional minimization. For the systems with irregular geometry, unusual boundary conditions, or heterogeneous composition, the finite element method provides a better solution compared to other numerical methods. A domain of interest is represented as an assembly of finite elements. Approximating functions in finite elements are determined in terms of nodal values of a physical field that is sought. A continuous physical problem is transformed into a discretized finite element problem with unknown nodal values. For a linear problem, a system of linear algebraic equations should be solved. Values inside finite elements can be recovered using nodal values.

Two features of the FEM are worth to be mentioned:

1) Piece-wise approximation of physical fields on finite elements provides good precision even with simple approximating functions (increasing the number of elements, we can achieve any precision).

2) Locality of approximation leads to sparse equation systems for a discretized problem. This helps to solve problems with a very large number of unknown nodal.

To summarize in general terms how the finite element method works, we list the main steps of the finite element solution procedure below.

1. Discretize the continuum. The first step is to divide a solution region into finite elements. The finite element mesh is typically generated by a preprocessor program. The description of mesh consists of several arrays many of which are nodal coordinates and element connectivities.

2. Select interpolation functions. Interpolation functions are used to interpolate the field variables over the element. Often, polynomials are selected as interpolation functions. The degree of the polynomial depends on the number of nodes assigned to the element.

3. Find the element properties. The matrix equation for the finite element should be established, which relates the nodal values of the unknown function to other parameters. For this task, different approaches can be used; the most convenient are: the variational approach and the Galerkin method.

4. Assemble the element equations. To find the global equation system for the whole solution region, we must assemble all the element equations. In other words, we must combine local element equations for all elements used for discretization. Element connectivities are used for the assembly process. Before the solution, boundary conditions (which are not accounted for in element equations) should be imposed.

5. Solve the global equation system. The finite element global equation system is typically sparse, symmetric, and positive definite. Direct and iterative methods can be used for the solution. The nodal values of the sought function are produced as a result of the solution.

6. Compute additional results. In many cases, we need to calculate additional parameters. For example, in mechanical problems, strains and stresses are of interest in addition to displacements, which are obtained after the solution of the global equation system.

Details about the FEM method have been discussed in another study [2]. Authors are suggested to follow other studies [2] for more information about FEM.

CONSENT FOR PUBLICATION

Not applicable.

CONFLICT OF INTEREST

The author declares no conflict of interest, financial or otherwise.

ACKNOWLEDGEMENT

Declared none.

REFERENCES

[1]   D. D. Moursund, C. S. Duris, *Elementary theory and applications of numerical analysis*, McGraw-Hill, New York, 1967.
[2]   S. C. Chapra, R. P. Canale, *Numerical methods for engineers*, McGraw-Hill Education, New York, USA.
[3]   M. K. Jain, S. R. K. Iyengar, and R. K. Jain, *Numerical methods for scientific and engineering computation*, Wiley Eastern Limited, India
[4]   S. S. Sastry, *Introductory methods of numerical analysis*, Prentice-Hall, India, 2003.
[5]   Taflove and S. Hagness, *Computational Electrodynamics: Finite Difference Time Domain Method*, Arttech House, USA.

EXERCISES

1. Calculate the potential difference using FD method on the rectangular trough considering inhomogeneous boundary conditions whose cross-section is shown below.

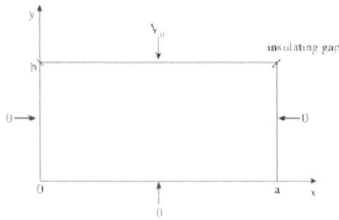

2. Determine the charge density ρ_L along a thin conducting wire (radius a, length $L(L \gg a)$) as shown below using MoM.

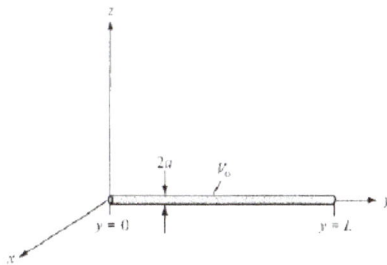

3. Use FDM to calculate the potentials at nodes 1 and 2 in the potential system as shown in the following figure.

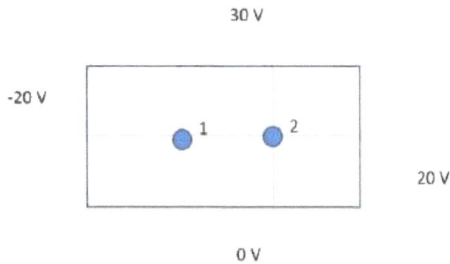

4. For a three-dimensional problem, the PML modification of Maxwell's equations yields 12 equations because all the six Cartesian field components split. Obtain the 12 resulting equations.

SUBJECT INDEX

www.ingramcontent.com/pod-product-compliance
Lightning Source LLC
Chambersburg PA
CBHW050820220326
41598CB00006B/270